典型阔叶红松林树木死亡动态与物种共存机制

Tree mortality dynamic and species coexistence mechanism in a typical mixed broadleaved-Korean pine forest

金光泽 朱宇 著

科学出版社

北京

内 容 简 介

本书针对森林群落物种共存与生物多样性维持机制研究中存在的问题和不足,以小兴安岭凉水典型阔叶红松林 9 hm² 大型森林动态监测样地为研究对象,采用广义线性混合模型及似然比检验,探讨了负密度制约对树木死亡的影响;应用广义地理加权回归模型及地理变异检验,探讨了生境异质性对树木死亡的影响;结合广义线性混合模型等手段,研究了树木死亡–生长种间权衡与物种共存之间的关系;利用负二项回归、零膨胀回归等模型手段,研究了树木死亡–更新种间权衡与物种共存之间的关系;根据物种共存理论对经典树木死亡模型的部分结构进行了修正,并集成森林动态监测样地数据,在微尺度下对树木死亡进行了动态模拟。

本书可供从事林学、生态建设与恢复,以及森林经理学的科研、教学、工程技术人员和相关专业高校学生参考。

图书在版编目(CIP)数据

典型阔叶红松林树木死亡动态与物种共存机制 / 金光泽,朱宇著. —北京:科学出版社,2020.2
ISBN 978-7-03-063356-9

Ⅰ. ①典… Ⅱ. ①金… ②朱… Ⅲ. ①阔叶树-红松-物种-生态环境-研究 Ⅳ. ①S791.247

中国版本图书馆CIP数据核字(2019)第256325号

责任编辑:张会格 刘 晶 / 责任校对:郑金红
责任印制:吴兆东 / 封面设计:刘新新

科学出版社 出版
北京东黄城根北街 16 号
邮政编码:100717
http://www.sciencep.com
北京盛通商印快线网络科技有限公司 印刷
科学出版社发行 各地新华书店经销

*

2020 年 2 月第 一 版   开本:720×1000 1/16
2020 年 2 月第一次印刷   印张:13
字数:260 000
定价:128.00 元
(如有印装质量问题,我社负责调换)

前　言

近年来,森林群落物种共存与生物多样性维持机制一直是群落生态学领域的研究热点。树木死亡以后不仅能为森林生态系统中其他个体提供资源,还能为其他树木的生长及幼苗的更新腾出空间并创造林窗条件,进而促进物种共存并维持森林群落生物多样性。阔叶红松(*Pinus koraiensis*)林是我国东北东部山区的地带性顶极植被,研究其树木死亡与物种共存之间的关系有助于人们更深入地理解阔叶红松林生物多样性维持机制。

目前,森林群落物种共存与生物多样性维持机制研究中存在诸多问题与不足。例如,生物邻体因子的计算采用了统一、固定的邻体影响半径,忽视了邻体半径种间变异的事实,这可能对负密度制约研究结果造成干扰;树木死亡经验模型多在个体尺度上建立,尺度较为单一,忽略了空间尺度变化对模型拟合结果的影响,而生境异质性对树木死亡的影响可能受空间尺度制约;关于树木死亡-生长种间权衡与物种共存关系的研究在热带森林开展较多,而在温带森林开展的此类研究鲜有报道;树木死亡与更新之间存在何种关系(如种间权衡关系)、是否与物种共存之间也存在关联等问题尚不十分明确;以往的树木死亡过程模型由于缺失树木的定位坐标、生境异质性信息和树木死亡动态监测数据,导致存在必须从裸地开始模拟、模型评价难度较大、某些模型结构比较粗糙等问题。

本书针对森林群落物种共存与生物多样性维持机制研究中存在的问题和不足,以小兴安岭凉水典型阔叶红松林 9 hm^2 大型森林动态监测样地为研究对象,采用广义线性混合模型及似然比检验、广义地理加权回归模型及地理变异检验,分别探讨了生物因子(负密度制约)与非生物因子(生境异质性)对树木死亡的影响;结合负二项回归、零膨胀回归等统计模型手段,探究局域因子(生物与非生物因子)对树木死亡、生长及更新影响的差异性,并探索了树木死亡-生长/更新种间权衡与物种共存之间的关系;根据物种共存理论对经典树木死亡模型的部分结构进行了修正,并集成森林动态监测样地数据,在微尺度下对树木死亡进行了动态模拟,以期为未来的研究提供一种新的思路。

本书承蒙国家自然科学基金重点项目(31730015)和国家自然科学基金面上项目(31870399)资助，特致诚挚谢意。感谢黑龙江凉水国家级自然保护区管理局的各位领导和职工在外业调查期间给予的大量帮助，感谢历届研究生在凉水 9 hm^2 森林动态监测样地外业调查中付出的辛劳。

由于作者水平有限，书中难免有不足之处，敬请读者不吝赐教。

作　者

2019 年 6 月于哈尔滨

目　　录

前言
1　绪论 ··· 1
　1.1　树木死亡 ·· 1
　　1.1.1　树木死亡概念 ··· 1
　　1.1.2　树木死亡驱动因子 ··· 1
　　1.1.3　树木死亡机制 ··· 4
　1.2　经典物种共存理论 ·· 5
　　1.2.1　负密度制约 ·· 8
　　1.2.2　生境过滤 ··· 9
　　1.2.3　生态位分化 ··· 10
　1.3　当代物种共存理论 ··· 11
　　1.3.1　生态位差异和平均适合度差异 ·· 11
　　1.3.2　稳定化机制和均等化机制 ·· 12
　1.4　树木死亡与物种共存的联系 ·· 13
　1.5　当前研究中存在的问题与不足 ··· 13
　1.6　研究目的与意义 ·· 14
　1.7　研究内容与技术路线 ·· 14
　　1.7.1　研究内容 ·· 14
　　1.7.2　技术路线 ·· 15
　1.8　拟解决的关键科学问题 ··· 15
2　研究区域概况与研究方法 ··· 16
　2.1　研究区域概况 ··· 16
　　2.1.1　地理条件 ·· 16
　　2.1.2　气候概况 ·· 17
　　2.1.3　土壤概况 ·· 17
　　2.1.4　植被类型 ·· 17
　2.2　数据收集 ·· 18
　　2.2.1　树木动态监测数据 ·· 18
　　2.2.2　地形和土壤状况 ··· 19
　　2.2.3　系统发育树和功能性状数据 ··· 20

2.3 主要分析方法 ··· 21
    2.3.1 统计模型 ··· 22
    2.3.2 过程模型 ··· 23
3 负密度制约对树木死亡的影响 ·· 24
  3.1 引言 ··· 24
  3.2 研究方法 ·· 25
    3.2.1 数据收集 ··· 25
    3.2.2 个体尺度树木死亡驱动因子的构建 ······························· 25
    3.2.3 个体尺度树木死亡模型 ··· 27
  3.3 结果 ··· 29
    3.3.1 最优变量组合 ·· 29
    3.3.2 个体尺度树木死亡的生物邻体驱动因子 ························· 30
    3.3.3 不同邻体半径下生物邻体效应种间变异检验 ··················· 31
  3.4 讨论 ··· 32
    3.4.1 生物邻体效应种间变异的影响因素 ······························ 32
    3.4.2 邻体半径对检验生物邻体效应种间变异的影响 ················ 33
    3.4.3 负密度制约导致的树木死亡 ······································· 34
  3.5 本章小结 ·· 35
4 生境异质性对树木死亡的影响 ·· 36
  4.1 引言 ··· 36
  4.2 研究方法 ·· 37
    4.2.1 数据收集 ··· 37
    4.2.2 样方尺度树木死亡驱动因子的构建 ······························· 37
    4.2.3 样方尺度树木死亡模型 ··· 37
  4.3 结果 ··· 41
    4.3.1 样方尺度树木死亡的地形驱动因子 ······························ 41
    4.3.2 不同空间尺度下地形效应空间变异检验 ························· 43
  4.4 讨论 ··· 46
    4.4.1 地形因子效应空间变异的影响因素 ······························ 46
    4.4.2 空间尺度对检验地形因子效应空间变异的影响 ················ 48
    4.4.3 生境异质性导致的树木死亡 ······································· 48
  4.5 本章小结 ·· 49
5 树木死亡-生长种间权衡与物种共存 ··· 50
  5.1 引言 ··· 50

5.2 研究方法 ·· 51
　　5.2.1 数据收集 ··· 51
　　5.2.2 局域驱动因子的构建 ·· 52
　　5.2.3 树木死亡与生长模型 ·· 52
　　5.2.4 树木死亡与生长关系检验 ··· 53
5.3 结果 ··· 53
　　5.3.1 局域因子对树木死亡与生长的影响 ···································· 53
　　5.3.2 树木死亡与生长的种内及种间关系 ···································· 55
5.4 讨论 ··· 56
　　5.4.1 局域因子对树木死亡与生长影响的异同 ······························· 56
　　5.4.2 树木死亡与生长的种内及种间关系和物种共存 ······················· 56
5.5 本章小结 ··· 59

6 树木死亡-更新种间权衡与物种共存

6.1 引言 ··· 60
6.2 研究方法 ·· 61
　　6.2.1 数据收集 ··· 61
　　6.2.2 样方尺度树木死亡与更新驱动因子的构建 ····························· 61
　　6.2.3 样方尺度树木死亡与更新模型 ·· 61
　　6.2.4 树木死亡与更新种间关系检验 ·· 62
6.3 结果 ··· 62
　　6.3.1 局域因子对树木死亡与更新的影响 ····································· 62
　　6.3.2 树木死亡与更新的种间关系 ··· 63
6.4 讨论 ··· 66
　　6.4.1 局域因子对树木死亡与更新影响的异同 ································ 66
　　6.4.2 树木死亡-更新种间权衡与物种共存 ···································· 67
6.5 本章小结 ··· 68

7 物种共存理论修正树木死亡过程模型

7.1 引言 ··· 69
7.2 研究方法 ·· 70
　　7.2.1 模型结构与计算公式 ·· 70
　　7.2.2 模型检验与参数敏感性分析 ··· 81
　　7.2.3 模拟程序设计与实现 ·· 82
7.3 结果 ··· 85
　　7.3.1 确证性检验 ··· 85
　　7.3.2 有效性检验 ··· 88
　　7.3.3 参数敏感性分析 ··· 90

7.4 讨论 …………………………………………………………………………… 92
    7.4.1 树木死亡空间格局 ……………………………………………………… 92
    7.4.2 模型结构、参数与输入的不确定性 …………………………………… 93
    7.4.3 森林动态监测大样地对树木死亡个体模型发展的支持 ……………… 95
    7.4.4 展望 ……………………………………………………………………… 95
7.5 本章小结 ………………………………………………………………………… 96
参考文献 ………………………………………………………………………………… 97
附录　树木死亡动态模拟(TMDS)模型程序核心源代码(C#) ………………… 110

1 绪 论

1.1 树木死亡

1.1.1 树木死亡概念

树木死亡是一个组织器官与环境之间的热力学平衡概念，它代表树木不再具有驱动能量梯度用于代谢或者更新的能力(董蕾和李吉跃 2013)。

1.1.2 树木死亡驱动因子

通常，树木死亡是多种因素共同作用的结果，其驱动因子一般可以分为生物因子和非生物因子(Franklin et al. 1987)。其中，由生物因子引起的树木死亡主要包括树木的自然衰亡、病虫害、种内与种间对空间及资源的竞争所导致的死亡(如负密度制约死亡) (Monserud et al. 2004, Das et al. 2012)；非生物因子引起的树木死亡主要包括干旱、洪水、火灾、风暴、极端温度、冰雪灾害、机械损伤所导致的死亡，同时还包括地形(海拔、坡度、坡向等)和土壤理化性质的空间异质性所导致的树木死亡。对树木生长不利的环境往往造成树木更高的死亡率(Wang et al. 2012)，而在不同的时间和空间尺度上，环境造成树木死亡的原因较多(Hamilton 1986)，这些死亡驱动因子反映了不同的生态过程。

1.1.2.1 资源竞争

生活在同一地区的两个树种，由于利用相同的资源(光照、水分、空间及营养物质)，常常导致种间竞争，其结果往往是两个树种的数量都下降，因为与其他树种进行竞争可以导致树种的死亡率增大。资源竞争不仅发生在不同物种的种群间，在同一物种的个体之间也会发生竞争。竞争排除原理认为，两个生态位完全相同的物种不能同时同地生活，其中一个物种最终必将另一个物种完全排除。由于同种或近缘物种(特别是同属种)常常在形态、生理、行为和生态方面相同或非常相似，因此种内竞争或近缘物种之间的竞争通常强于远缘物种之间的竞争。在森林经理学林木生长与收获模型研究中，单木竞争因子是单木枯损经验模型中一项必不可少的自变量(Hamilton 1986)；而在 CROBAS 碳平衡过程模型中，竞争也同样被认为是导致林分枯损的一项重要因素。张泽浦等(2000)利用树木邻体竞争指数分析竞争影响日本落叶松(*Larix leptolepis*)种群个体生长率和死亡率的影响时发现，引入竞争指数可以较好地预测个体在竞争压力下的死亡概率，但不能明显改善对生长速率的预测效果。

1.1.2.2　树木衰老

Lugo 和 Scatena(1996)在波多黎各 Luquillo 热带雨林研究树木死亡时发现，随着树木胸径增大，其年龄逐渐增长，当树木接近该树种最大寿命时，由于树木光合作用等生理功能的衰退，树木的死亡概率会大大增加。Coomes 和 Allen(2007) 对新西兰山毛榉 (*Nothofagus solandri*) 天然异龄林进行了 19 年的长期动态监测发现，随着径级的增大，树木死亡呈现出"U"形变化趋势，这表明衰老是树木死亡的原因之一。Wang 等(2012)对长白山阔叶红松林 25 hm^2 动态样地的研究发现，对于小树和中树，初始胸径大小对其存活存在显著的正效应，即随着胸径的增大，小树和中树死亡率降低，而对于老树，初始胸径对其存活存在不显著的负效应，即树木初始胸径大小不再促进树木存活。Wu 等(2017)在八大公山亚热带常绿落叶阔叶混交林 25 hm^2 动态样地的研究也得到了相似的结论，对于小树和中树，初始胸径对其存活存在显著的正效应，而对于老树，初始胸径对其存活存在不显著的负效应。以上研究表明，随着年龄的增大并接近该树种寿命极限时，由于生理功能衰退，树木的死亡率逐渐增大。

1.1.2.3　虫害及病原菌传播

陈志成和万贤崇(2016)在虫害叶损失造成的树木非结构性碳减少与树木生长、死亡的关系研究进展中总结，虫害或虫害与其他环境胁迫(如干旱)并发造成的叶损失可导致树木死亡(Kosola et al. 2001, Gaylord et al. 2013, Jacquet et al. 2014)。虫害导致的叶损失会减少树木光合作用产物，增加非结构性碳(nonstructural carbohydrate, NSC)消耗，从而使树木体内碳储备降低(Anderegg and Callaway 2012, Anderegg et al. 2015)，生长减弱，当 NSC 储备降低到一定阈值时，树木会由于碳饥饿而死亡。此外，Janzen-Connell 假说认为病原菌的传播容易造成同种邻体幼苗的死亡，由于有害生物从母株传播到附近的更新后代上，聚集的寄主有利于专一性食草动物和病原菌的繁殖与传播，从而受到有害生物侵害，使得同种邻体表现出较高的死亡率(Janzen 1970, Connell 1971)。

1.1.2.4　干旱

干旱是对植物影响最大的非生物胁迫之一，它可以影响树木的一系列生理过程，进而影响树木的生长和存活(Barigah et al. 2013)。随着全球气候变化，气温升高、降水格局的改变，干旱发生的普遍性越来越强(Galle et al. 2010, Hartmann 2011, Choat 2013)，特别是在干旱、半干旱地区，干旱发生的频率呈现增加的趋势(Ryan 2011)。Greenwood 等(2017)对全球 58 个由干旱导致的树木死亡的研究

(遍布欧洲、美洲、亚洲和非洲)进行整合分析发现,树木死亡随着干旱程度的增大而加剧,这一结果具有全球一致性和普适性;Peng 等(2011)对加拿大北方林的研究结果表明,干旱是该地区树木死亡率上升的主要原因;Phillips 等(2009)在南美洲亚马孙热带雨林的研究结果发现,由于水分胁迫导致该地区森林树木的生长十分脆弱;Mantgem 等(2009)在美国西部森林的研究结果表明,干旱可能是树木死亡率升高的重要因素。

1.1.2.5 森林火灾

森林火灾对林木树冠、树干和树根造成的伤害,以及对后续的病虫害侵袭和气候的影响是长久的,因此森林火灾造成的林木死亡现象会持续多年。雷击是引发森林火灾的主要自然火源(杜春英等 2010)。Grayson 等(2017)使用 FOFEM(V5)模型对美国西北部太平洋沿岸地区 14 个针叶树种树木火后死亡概率进行了预测,在该模型中使用树冠烧焦体积和基于物种与胸径的树皮厚度作为预测树木死亡的重要变量(Woolley et al. 2012, Grayson et al. 2017),林木死亡率随着树冠烧死百分比的增大而增加,而随着树皮厚度的增大而减少;树木死亡率受树冠烧伤和树干烧焦严重程度相互作用的影响(蔡慧颖 2012)。林火发生时,直接烧伤和烧死森林植物,尤其是森林中的草本植物、灌木和幼树,高强度火也可以烧伤和烧死高大的树木。火灾发生后导致森林环境的改变,间接影响到树木的生长和生存,其中一些树木生长受到抑制(李俊清等 2017)。

1.1.2.6 风暴

超过 10 m/s 的大风对树木有破坏作用,产生机械伤害,如造成风折、风倒、风拔等。13~16 m/s 的风速,可以使每平方米树冠表面积受到 15~20 kg 的压力。强风作用下,根系生长浅的树种(如生长在小兴安岭谷地永冻土或季节性冻土上的云杉和冷杉)能被连根吹倒,形成风倒木(windthrow)。感染病虫害、生长衰弱、老龄过熟的林木,被强风吹折树干,称为风折,如干中折、干基折等。强热带风暴(或台风)登陆、地方性的暴风雨是引起树木风折、风倒的重要原因之一。老龄大树树倒是森林内部干扰的主要形式,风加剧了这一过程,这是原始林树木死亡和形成林窗的主要原因(李俊清等 2017)。在 2008~2009 年,黑龙江凉水国家级自然保护区遭受到了严重的风暴侵袭,造成了大量的风倒木,尤其是大径级的红松(Zhen et al. 2013);Csilléry 等(2017)对位于阿尔卑斯山西部地区和汝拉山区的 115 个针叶混交林林分研究发现,风暴强度增强会增加上述森林树木的死亡风险。

1.1.2.7 冰雪灾害

温度下降,达不到树种最小有效积温,会导致树木生长受到抑制,致使树木死亡率增大。温度低于一定的数值，生物便会因低温而受害。低温对生物的伤害可分为冷害、冻害和霜害三种。冷害是指喜温生物在 0℃以上的温度条件下受害或死亡。冻害是指冰点以下的低温使生物体内(细胞内和细胞间隙)形成冰晶而造成的损害(李俊清等 2017)。金毅等(2015)研究了浙江省古田山亚热带常绿阔叶林 24 hm^2 动态监测样地群落结构和组成动态，并且探讨了 2008 年冰雪灾害的影响发现，群落整体补员不足且死亡率较高，说明 2008 年发生的冰雪灾害对群落的短期动态产生了较大的负面影响，较高的死亡率导致该森林群落绝大多数径级个体数量减少，且群落在多个粒度上呈广泛的衰退。此外，冰雪灾害造成大树枝干折断等物理性破坏和遮挡阳光，因此在短期内导致大量幼苗死亡，可能进一步加剧小径级的补员不足(Darwin et al. 2004, Vowels 2012)。冰雪灾害对单株树的伤害程度，从仅损坏一些嫩枝、使茎弯曲至地面，到部分树冠受损，甚至使树干完全破裂(曹坤芳和常杰 2010)。冰积累对树木有严重的物理伤害。冰积累的厚度达 0.6～1.3 cm 时会使小枝或有缺陷的树枝脱落，1.3～2.5 cm 时会产生明显的破损(Lemon 1961)。当冰覆盖在过分伸展树枝的倾斜幼树时，在其树皮光滑的外表上会形成硬结和很明显的损伤。2008 年受拉尼娜现象(即反厄尔尼诺现象)及大气环流异常等的影响，北方强烈的冷气流南下与南方暖湿气流在我国南方交汇，南方诸省遭受大范围的特大冰雪灾害。冰雨在树木等物体上凝结成冰。直径 1 cm 左右的树枝上结冰的厚度可超过 10 cm，冰重量可达其自身重量的数十倍，导致树木折干、倒伏、翻蔸、弯斜、断枝、断冠、劈裂、爆裂等，促使林木受到破坏而死亡(曹坤芳和常杰 2010)。

1.1.3 树木死亡机制

树木死亡受生物和非生物因子共同影响，而碳饥饿(carbon starvation)和水力失衡(hydraulic failure)假说可能是树木死亡的两个潜在影响机制。生物邻体间相互作用时发生的种内种间资源竞争，尤其是同种或近缘种间光资源的竞争，导致某些树木得不到充足的光照，进而造成光合作用能力下降，限制碳固定，并最终引发树木因碳饥饿而死亡。碳饥饿假说最早在 1975 年被提出(Parker and Patton 1975)，但长期以来，碳饥饿多用来解释干旱胁迫造成的树木死亡。Bossel(1986)和 Mueller-Dumbois(1987)认为长期的干旱胁迫会打破植物碳摄取与碳支出的平衡，从而使植物死亡；McDowell 等(2008)认为干旱导致气孔关闭，阻止水分进一步散失，但是同时也导致了光合作用的碳摄取减少，而植物新陈代谢对 NSC 的继续消耗就导致了碳饥饿的产生；虫害啃食叶片直接导致光合碳固定降低；遮阴会限制光合作用，也会导致碳固定减少(Boardman 1977, McDowell et al. 2008, McDowell 2011, Sevanto et al. 2014, Anderegg et al. 2015)；Sevanto 等(2014)也证明了遮阴能导致树木幼苗因碳饥饿死亡。叶片是光合作用器官，虫害啃食叶片造成树木叶损失，不但会导致整棵树叶片中 NSC 的损耗，还会减少碳的净获取，另外树木叶片重新萌发也会增加 NSC 消耗。总之，因虫害导致的叶损失使树木光合作用碳获取减少和 NSC 消耗增加，致使树木体内碳储备降低(Li et al. 2002, Anderegg and Callaway 2012, Anderegg et al. 2015)，从而使树木生长活力减弱，当 NSC 储备降低到一定阈值时树木会因碳饥饿而死亡。此外，随着树木胸径增大、年龄增高，树木逐渐衰老，当树木接近其所属树种最大寿命时，由于树木生理功能的衰退，树木的死亡率也会大大增加，最终导致树木因碳饥饿而死亡。

生境异质性与物种生境偏好共同作用所导致的生境过滤效应则会使得某些树木因生长在不适宜的生境中而得不到充足的水分、光照，最终导致水力失衡、碳饥饿而死亡。例如，若喜光树种生长在荫蔽的生境、湿生树种生长在干旱的生境，则这些树木都极有可能因生境过滤而死亡。水力失衡也是干旱导致树木死亡的主要生理学机制(McDowell et al. 2008)。水力失衡是指减少的土壤水分供应和高的蒸发需求导致木质部导管和根系产生空穴化，或干旱导致树木的蒸腾超过临界蒸腾，枝条导水率接近于零，植物长距离水分运输受限，使树木发生不可逆的干化现象(McDowell et al. 2008)。严重的干旱可使气孔完全关闭，使光合碳同化降低到零(McDowell et al. 2008)。另外，在干旱时的水分限制会阻碍韧皮部对碳水化合物的运输，也会导致碳饥饿的发生(Sala et al. 2010)。在干旱胁迫情况下，栓塞(cavitation embolism)的积累将导致木质部水分运输失调，甚至使树木死亡(Brodribb and Cochard 2009)。另外，由于水力失衡而出现的木质部栓塞和空穴会进一步加剧水分运输障碍，而修复空穴则需要大量 NSC，这使植物陷入两难选择(Tyree and Dixon 1986)。

1.2 经典物种共存理论

群落生态学家们一直试图解释为何热带雨林有如此高的物种多样性，为什么最有竞争能力的物种的多度不能无限增加至排除掉其他树种(Hubbell 1979, Wright et al. 1999)。一般认为，各物种功能性状的取舍或权衡(trade-off)和组合的不同决定了其生活史对策(如资源利用方式)的不同，也决定了各物种在群落中所占有的生态位不同，进而决定了多物种的稳定共存(Hutchinson 1957, Vandermeer 1972, Silvertown 2004)。生态位机制主要包含生物互作(biotic interaction)(生物因子)(Gause 1934)及生境过滤(environmental filtering/habitat filtering)(非生物因子)(Tilman 2004)两种机制。Diamond 于 1975 年首次提出了"群落构建"(community assembly, 也可译为"群落组配")的概念，群落构建是群落内物种组成与配置的规律，指的是群落物种多样性的形成和维持(Chesson 2000)，本质上是探讨群落内多物种共存的机理，以及共存的影响因素和规律。群落如何构建(或组配)是群落生态学中的核心问题之一,生物互作和生境过滤被认为是解释群落构建与群落生物多样性维持的两个重要机制。目前物种共存理论是群落构建当中的重要一环(图1-1)(HilleRisLambers et al. 2012)。

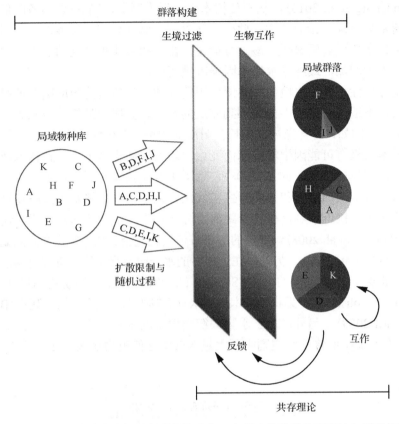

图1-1 "透过共存理论镜头重新思考群落构建"一文提出的物种共存理论与群落构建之间的关系。生物互作与环境过滤是物种共存的两个重要机理(HilleRisLambers et al. 2012)

Fig. 1-1 Species coexistence and community assembly mechanisms from the study of "Rethinking community assembly through the lens of coexistence theory". Biotic interactions and abiotic environmental filtering are the most important mechanisms fostering species coexistence

当前大致存在两类不同的视角探讨森林群落物种共存和群落构建机制。第一类视角为基于格局推断森林物种共存和群落构建过程,该视角按照方法论可分为三个方面。①该方面包括两个内容。一是基于功能性状的分布格局来探讨生物互作与生境过滤在群落构建过程中的相对作用,即如果共存物种的性状分布相对于零模型表现为聚集的格局,那么推断生境过滤是群落构建的主要驱动力;反之,如果性状分布表现为发散的格局,那么物种之间的生物互作(竞争排斥)则起着主导作用(Ackerly and Cornwell 2007)。二是基于亲缘关系的谱系群落生态学,与基于功能性状的方法类似，如果生境过滤作用占主导地位，那么相似生境筛选出适应能力相似、亲缘关系较近的物种，从而表现为谱系的聚集；相反，生物互作(竞争排斥)会使得生态位相似的物种无法稳定共存于同一环境，导致群落内物种的亲缘关系则较远，表现为谱系的发散(Webb et al. 2002)。这里是以植物功能性状具有系统发育保守性为前提提出的假设(Webb et al. 2002)。例如，Swenson 等(2012)利用净亲缘关系指数(net relatedness index, NRI)及标准化后的平均功能性状距离(standardized effect size of the mean pairwise trait distance, SES.PW)指数，结合系统发育和功能性状数据分析了全球多个温带和热带大型森林动态样地(BCI, Luquillo, Wabikon Lake, Smithsonian Conservation Biology Institute, 古田山)的谱系和功能性状多样性扩散格局，发现生境过滤在群落构建中扮演着十分关键的角色；任思远等(2014)利用净亲缘关系指数分析了河南省宝天曼暖温带-北亚热带过渡带落叶阔叶林 1 hm^2 动态样地不同径级系统发育结构的变化，结果表明生物互作(密度制约)是该地区森林群落物种多样性维持的重要机制。②通过空间点格局分析(spatial point pattern analysis, PPA；包括单变量和双变量)检验物种之间、物种与生境之间的关联性(正或负，即吸引或排斥)，并由此推断生物互作(如密度制约)和生境过滤过程的重要性(Ripley 1977, Diggle 2003)。例如，张觅等(2014)采用空间点格局分析对小兴安岭凉水谷地云冷杉林 9.12 hm^2 动态监测样地的群落组成和空间格局进行分析，研究结果表明，优势种冷杉幼树 I[胸径(DBH)：1~5 cm]和幼树 II(DBH: 5~10 cm)与成年树(DBH: >10 cm)利用生境的方式均不同，红皮云杉幼树 I 与成年树利用生境的方式不同，而幼树 II 与成年树利用生境的方式相同；冷杉与红皮云杉的幼树 I 与幼树 II 均在 1~50 m 尺度上呈现相对于成年树额外的聚集，在尺度>5 m 时，随着径级增大，这种额外的聚集逐渐减小，说明密度制约效应在起作用。③利用多元回归树(multivariate regression tree, MRT)(De'ath 2002)对生境类型进行分类，并结合 Torus 转换检验法(torus-translation test)(Harms et al. 2001)检验物种与生境之间的关联性，由此推断生物互作和生境过滤过程。例如，Lai 等(2009)采用多元回归树和 Torus 转换检验法对古田山亚热带常绿阔叶林 24 hm^2 动态监测样地的物种与生境关联性进行分析，研究结果表明，物种与生境关联性在不同的生活史阶段存在差异，大多数物种幼树和中树的生境偏好保持一致，但是在老树阶段发生改变。Liu 等(2018)采用多元回归树与 Torus 转换检验法对小兴安岭丰林 30 hm^2 大型监测样地的物种与生境关联性进行分析，研究结果表明，样地中大多数树种均存在生境偏好，说明生态位理论在生物多样性维持中扮演重要角色。但在不同生活史阶段，大多数树种与生境之间不存在一致性的联系。总之，第一类探讨森林群落物种共存理论和群落构建机制的视角中的格局是指静态格局。

第二类视角为基于森林动态监测推断物种共存和群落构建过程：通常采用广义线性混合模型，并与树木死亡、生长、更新等动态监测数据相结合，尤其是树木死亡动态数据。生物与非生物自变量参数估计值的大小、正负和显著性对应着生物互作与生境过滤等物种共存理论的解释。第二类视角采用森林动态数据，这能够充分发挥大型固定森林动态监测样地的特色和优势。尤其是近年来，利用大型森林动态监测样地的树木死亡动态数据开展关于密度制约、生境过滤等物种共存理论与生物多样性维持机制的研究在国内外得到了蓬勃发展。Comita 等(2010)在《科学》(Science)期刊上发表的关于利用热带森林幼苗死亡动态数据，探讨同种与异种密度制约效应的论文并引起广泛关注，研究结果表明，密度制约效应，尤其是同种邻体负密度制约效应是重要且广泛存在的森林生物多样性维持机制。

经典物种共存理论强调具体的物种共存理论机制，比较常见的有负密度制约、生境过滤和生态位分化等。

1.2.1 负密度制约

通常情况下，生物邻体互作可以指负密度制约，它是物种共存及生物多样性维持机制之一(Chesson 2000, HilleRisLambers et al. 2012)。Janzen 和 Connell 分别于 1970 年和 1971 年撰文，提出了著名的物种共存理论之一——负密度制约(negative density dependence, NDD)假说，也被称为 Janzen-Connell 假说(Janzen 1970, Connell 1971)。该假说认为，种子扩散以母株为中心，邻体母株的种子和幼苗存在较高的死亡率，这有助于维持热带森林生物多样性。有害生物从母株传播到附近的更新后代(recruitment offspring)上，聚集的宿主有利于专一性食草动物和病原菌的繁殖与传播，由于受到有害生物侵害，使得同种邻体表现出较低的生长率和存活率。遭受侵害的物种，其同种邻体间距增大，将会给与它天敌不同但有相似资源需求的其他物种腾出空间和资源，从而促进物种共存，并提高热带森林的物种多样性。Comita 等(2014)收集了从 Janzen-Connell 假说提出后 40 多年间(1970～2013 年)的共 1038 篇研究文献和案例(包括遍布于欧洲、亚洲、美洲、非洲的温带和热带森林)，并基于此进行整合分析(meta analysis)，用以评价 Janzen-Connell 假说中关于负密度制约和负频率制约的证据充分性，研究结果为 Janzen-Connell 假说中关于种子和幼苗负密度制约和负频率制约死亡在世界范围植物群落内的广泛性提供了充分的证据，说明虫害和病原菌传播是森林群落中幼苗和种子死亡的重要驱动因子。由于同种邻体或亲缘关系较近或功能性状相似的邻体之间具有相同或相似的资源需求，从而导致强烈的种内或种间竞争(Burns and Strauss 2011)，树木的生长效率由此降低，出现邻体互补效应(Chen et al. 2016)；树木死亡率升高，并导致同种负密度制约(conspecific negative density dependence, CNDD)死亡和系统发育负密度制约(phylogenetic negative density dependence, PNDD)死亡。Zhu 等

(2015b)在巴拿马 Barro Colorado Island (BCI)热带雨林 50 hm² 动态监测样地，利用多期树木死亡监测数据，采用广义线性混合模型检验 3 种生物邻体因子对树木存活的相对重要性进行研究时发现：同种邻体密度对幼树和中树的存活具有显著的负效应，即 CNDD，而对大树的存活具有边缘显著的、微弱的正效应；异种邻体密度对树木存活的效应随生活史阶段不同并无明显的变化趋势，且强度弱于同种邻体密度，表明在 BCI 热带雨林群落中，种内资源竞争的强度大于种间竞争，而种内竞争大多发生在树木较早生活史阶段。Wang 等(2012)在长白山阔叶红松林 25 hm² 动态监测样地，利用树木死亡动态监测数据，采用广义线性混合模型，分析林木大小、生物邻体因子、地形和土壤因子对树木存活的相对重要性时发现：在群落水平上，总的生物邻体胸高断面积和同种邻体占总邻体比例对树木存活存在显著的负效应，即 CNDD，表明种内、种间资源竞争是树木死亡的重要因素。Wu 等(2017)在湖南省八大公山亚热带常绿落叶阔叶混交林 25 hm² 动态监测样地，以样地内的枯立木为研究对象，采用空间点格局分析和广义线性混合模型检验林木大小、生物邻体和地形因子对枯立木发生的相对贡献时发现：无论在群落水平、同资源种团水平还是物种水平上，同种邻体胸高断面积和异种邻体胸高断面积指数与枯立木发生均存在显著的正相关关系，即 CNDD 与异种邻体负密度制约(heterospecific negative density dependence, HNDD)，表明该群落存在较强的种内和种间竞争，并导致自我稀疏，生物因子对枯立木发生的相对贡献大于地形因子。Fan 等(2017)在吉林省蛟河阔叶红松林 21.12 hm² 动态监测样地，利用树木死亡动态监测数据，采用广义线性混合模型检验林木大小、同种和异种邻体胸高断面积、地形、多度级和生长型对树木存活的相对重要性时发现：在群落水平以及几乎所有同资源种团水平上，同种邻体胸高断面积对树木存活均存在显著的负效应，即 CNDD。该研究结果发现，对于树木死亡而言，种内资源竞争比种间竞争起到了更重要的作用。但是，当前负密度制约研究中的生物邻体因子计算往往忽略了邻体半径种间变异的重要性。

1.2.2 生境过滤

生境过滤是物种共存及生物多样性维持的机制之一(Chesson 2000, HilleRisLambers et al. 2012)。生境过滤与生物互作是一对相反的作用力(Webb et al. 2002)，生境过滤主要由生境异质性与物种生境偏好共同作用所引发，使得共存的物种利用相似的资源，而生境因子一般包括地形(海拔、坡度、坡向、凹凸度、山体阴影等)和土壤理化性质等。地形的变化引起水、肥、气、热的重新分配，属于间接生态因子。在山地条件下，地形条件是影响生物生长发育的重要因素(韩大校和金光泽 2017，李俊清等 2017)。地形和土壤理化性质存在一定的相关性，地形可以对土壤理化性质的空间分布进行重新分配。由于每个树种自身的生活史策略(life history strategy)(例如，对光强和水分的适应策略、耐阴性和耐旱性)使其具有特定的生境偏好，所以几乎不可能使所有物种都出现在所有类型的生境，生境过滤将淘汰掉那些因为得不到充足的水分和光照而不能正常生长在某类生境的树木，并将腾出的空间和资源给适宜在此类生境中生长和存活的其他物种，生态位相似的物种组共存于某类特定生境(赵雪等 2015)，从而促进物种共存并维持森林的物种多样性(Hutchinson 1957, Baldeck et al. 2017)。生境异质性会导致某些树木因无法得到充足的光照和水分而死亡。生境异质性和特定树种的生境偏好能够引发树种空间聚集格局，使得特定树种被限制在一定空间范围内生长(Piao et al. 2013)，这在以往的温带和热带森林树种空间分布特征的研究中多有证实，如西双版纳热带雨林、小兴安岭凉水典型阔叶红松林及巴拿马 BCI 热带雨林、泰国 Huai Kha Khaeng(HKK)热带雨林和马来西亚 Pasoh 热带雨林等大型森林动态监测样地(Condit et al. 2010, Lan et al. 2011, Lan et al. 2012)。徐丽娜和金光泽(2012)发现小兴安岭凉水典型阔叶红松林主要树种的空间分布呈现聚集态势，并且树种的空间分布格局与地形因子密切相关。Lan 等(2012)发现西双版纳热带雨林 20 个主要树种当中，有 14 个树种的空间分布与某一个或两个地形因子显著相关，表明生态位控制着森林群落中各个物种的空间分布。若干温带与热带的森林动态监测样地，如长白山温带森林及厄瓜多尔 Yasuní 低地热带雨林动态监测样地研究报道显示，树种的空间分布格局、树木生长和存活与其所处生境息息相关(Zhang et al. 2011, Metz 2012)。但迄今为止，极少有在多尺度下探究生境异质性与树木死亡之间关系的报道。

1.2.3 生态位分化

除了负密度制约与生境过滤可促进物种共存之外，资源生态位分化(resource niche partitioning)也可促进森林群落物种共存并维持生物多样性。生态位分化是经典物种共存理论的精髓。物种在群落中的共存以生态位的分化为前提，生态位相同的物种可能因竞争共同的资源而发生竞争排除，不能稳定共存(Vandermeer 1972)。特别是近年来开展的关于树木死亡-生长种间权衡关系的研究表明，通过生态位分化可促进物种共存并维持森林生物多样性(Kitajima and Poorter 2008)。竞争排除原理认为，若两个物种实现了共存，则在它们之间必然会存在生态位差异，它们的生态位必然不可能完全重叠。森林群落中的不同树种通常具有不同的生态对策，r 对策树种具有生长速度快、寿命短、死亡率较高等特征。r 对策树种将资源主要投入于繁殖和生长能力，而这种取舍导致投资于竞争能力及存活能力的资源不足。相反，群落中 K 对策树种，生长速度缓慢但寿命长，具有较高的存活率。K 对策树种将资源主要投入于竞争及存活能力，而导致投入于繁殖和快速生长能力的资源不足。不同生态对策的树种占据不同的生态位，因此，资源生态位的分化有利于促进物种共存并维持森林群落生物多样性。Wright 等(2010)在巴拿马 BCI 热带森林 50 hm^2 动态监测样地的研究发现，树种在有利条件下的快速生长与不利条件下的低死亡率之间存在显著的种间权衡关系并促进物种共存，但这种权衡关系只存在于小树之间，而大树之间不存在这种显著的权衡关系；Iida 等(2014b)在马来西亚 Pasoh 热带森林 50 hm^2 动态监测样地的研究中同样发现，树种相对生长率与死亡率之间存在显著的权衡关系并促进物种共存，但这种权衡关系也只存在于小树阶段。目前在温带森林开展的此类研究却鲜有报道。

1.3 当代物种共存理论

当代物种共存理论是经典物种共存理论的补充和完善。经典物种共存理论强调具体的物种共存机制，例如，树种对光、土壤等资源的分化利用，物种的时间生态位分化和空间生态位分化等。当代物种共存理论不关注具体的共存机制，而是主要集中于揭示影响物种共存的一般性规律。Chesson(2000)提出的关于物种共存的综合性理论框架，即"当代物种共存理论"。

在当代物种共存理论框架下，物种间的差异被划分为两类综合性的抽象差异——生态位差异和平均适合度差异。前者促进物种共存，对应稳定化机制；后者导致竞争排除，对应均等化机制(图 1-2)。

1.3.1 生态位差异和平均适合度差异

相对于中性理论中假定物种功能等同，当代物种共存理论将种间的差异分为两类：生态位差异(niche difference, ND)和平均适合度差异(average fitness difference, AFD)(图 1-2)。生态位差异和平均适合度差异这两个非常抽象的概念，均不涉及具体的物种差异，而是对真实的物种差异进行高度概括与抽象。对于生态位差异，我们可以作如下理解：深根性树种与浅根性树种通过根系的长短实现对资源的分化利用；对于平均适合度差异，类似地可以将其设想为物种对相同资源在利用效率上的差异(HilleRisLambers et al. 2012)。由此可见，当代物种共存理论中的生态位差异包括了经典物种共存理论的精髓，即生态位分化。生态位差异越大，越有利于稳定共存，而平均适合度差异越大则越有利于竞争排除。对于任何一个现实的群落，生态位差异和平均适合度差异的相对大小决定了物种是稳定共存还是会发生竞争排除，这是当代物种共存理论最为核心的内容。当物种既无生态位差异也无适合度差异的时候，群落即为中性格局。从这个意义上来讲，中性理论和当代物种共存理论本质上是不冲突的。

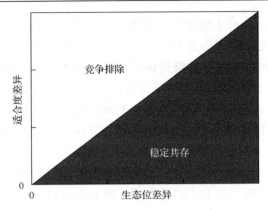

图 1-2 当代物种共存理论框架。物种间差异分为生态位差异和平均适合度差异,前者促进物种共存,后者利于竞争排除。物种能否稳定共存取决于生态位差异和平均适合度差异的相对大小:灰色区域,生态位差异大于平均适合度差异——稳定共存;白色区域,平均适合度差异大于生态位差异——竞争排除。生态位差异对应稳定化机制,平均适合度差异对应均等化机制(储诚进等 2017)

Fig. 1-2 The conceptual diagram of contemporary coexistence theory. Species differences are categorized into two groups: niche differences and average fitness differences. Niche differences maintain species stable coexistence, and average fitness differences drive competitive exclusion. For a given community, the balance between niche differences and average fitness differences determines the outcome of competition. In the grey region, niche differences are larger than average fitness differences, which results in stable coexistence. In the white region, average fitness differences are stronger than niche differences, which results in competitive exclusion. Niche differences correspond to stabilizing mechanisms, and average fitness differences correspond to equalizing mechanisms

1.3.2 稳定化机制和均等化机制

物种的生态位差异导致了稳定化机制(stabilizing),平均适合度差异导致了均等化机制(equalizing)。如上所述,在当代物种共存理论当中,稳定化机制与均等化机制共同决定了物种能否实现稳定共存(Chesson 2000):稳定化机制通过物种之间的生态位分化减弱或者消除竞争的影响,而均等化机制通过降低物种之间的适合度差异来促进共存。

稳定化机制涵盖了大部分经典的有关物种共存的假设和理论,包括种间资源生态位分化(Vandermeer 1972)、Janzen-Connell 假说中物种特异性天敌(Connell 1971, Barigah et al. 2013)等。均等化机制是通过降低物种之间的适合度差异或者物种在竞争能力方面的差异来促进共存的。以对资源的竞争为例,物种对资源的竞争能力可以通过 R^* 值来表示,即某物种在群落中能够维持下去所需的最低资源水平。当多个物种同时利用相同的限制性资源时,具有最低 R^* 值的物种将竞争排除掉所有其他的物种。因此,任何能减少物种之间 R^* 值差异的因子都可被归为均等化的机制,如对竞争优势种的部分去除等。不同于稳定化机制,均等化机制不会导致负密度制约。

经典物种共存理论的工作几乎都是以稳定化机制为研究对象的,强调物种之间的生态位分化,却很少关注均等化机制。树木死亡-生长种间权衡可通过稳定化机制(生态位分化)和均等化机制(适合度趋同)共同实现物种稳定共存。

1.4 树木死亡与物种共存的联系

并非所有的树木死亡均对森林产生负面效应,某些条件下的树木死亡对于促进森林群落物种共存及维持森林生物多样性具有重要的正面效应。之所以可以利用树木死亡动态数据来探讨物种共存机理,是因为树木死亡影响因素及其死亡机理(碳饥饿、水力失衡假说)与物种共存理论(负密度制约、生境异质性)之间存在着密切的联系。另外,相对于树木生长,树木死亡为竞争的存在提供了更强有力的证据(Gray and He 2009)。生物邻体间相互作用发生的种内、种间竞争,尤其是同种或近缘物种之间光资源的竞争,导致某些喜光树种树木得不到充足的光照,光合能力下降,限制了固碳能力,使树木因碳饥饿而死亡。另外,由于密度制约引起的同种邻体间病虫害暴发、啃食叶片,直接导致 NSC 损耗,树木可能因碳饥饿而死亡。生境异质性及物种生境偏好导致的生境过滤则会使得某些湿生树种或喜光树种树木因生长在不适宜的生境(如干旱生境或荫庇生境)中而得不到充足的水分或光照,进而导致水力失衡或碳饥饿而死亡。例如,喜光树种若生长在荫庇的生境、湿生树种生长在干旱的生境,都极有可能因生境过滤(碳饥饿或水力失衡)而死亡。以往关于物种共存理论的研究优先采用树木死亡动态数据(Wang et al. 2012, Zhu et al. 2015b, Wu et al. 2017)。此外,森林动态模拟研究中的树木死亡过程模型与物种共存理论之间有诸多可相互借鉴,但迄今为止尚无此类研究报道。

1.5 当前研究中存在的问题与不足

当前研究存在诸多问题与不足:①生物邻体因子的计算采用了统一、固定的邻体影响半径,忽视了邻体半径种间变异的事实,这可能对负密度制约研究结果造成干扰;②树木死亡经验模型多在个体尺度上建立,尺度较为单一,忽略了空间尺度变化对模型拟合结果的影响,而生境异质性对树木死亡的影响可能受空间尺度制约;③关于树木死亡-生长种间权衡与物种共存关系的研究在热带森林开展较多,而在温带森林开展的此类研究鲜有报道;④树木死亡与更新之间存在何种关系(如种间权衡关系)、是否与物种共存之间也存在关联这些问题尚不十分明确;⑤以往的树木死亡过程模型由于缺失树木的定位坐标、生境异质性信息和树木死亡动态监测数据，导致存在必须从裸地开始模拟、模型评价难度较大、某些模型结构比较粗糙等问题(Botkin et al. 1972, Leemans and Prentice 1987, Busing 1991, 桑卫国和李景文 1998, Yan and Shugart 2005, 国庆喜和葛剑平 2007)，而森林动态监测样地和物种共存理论的发展可为树木死亡过程模型结构的修正及模型的验证提供支撑。当前研究存在问题与不足详见 3.1 节、4.1 节、5.1 节、6.1 节、7.1 节。

1.6 研究目的与意义

原始阔叶红松(*Pinus koraiensis*)林是我国极为珍贵的森林资源，是欧亚大陆北温带最丰富、最多样的森林生态系统，蕴藏着极为丰富的动植物资源，对我国东北地区的生态安全起着巨大的防护作用(李景文 1997)。我国阔叶红松林主要分布在东北地区的东部山区，属于大陆性季风气候区。该区生长季虽短，但水热同步，气候湿润，为森林植物生长发育提供了适宜而又优越的生长环境，构成了北温带的地带性森林植被——以红松为建群种的针阔混交林。该地区是我国主要的天然林区之一，在过去的数十年间，该林区为国家提供了大量的木材，有力地支援了我国的经济建设。但是该地区目前的客观形势不容乐观，百年前还是浩瀚的林海，近一个多世纪内森林面积和蓄积急剧地减少，目前我国仅有的阔叶红松林已经陷入"两危一困"的境地，即森林资源危机、森林生态危机和经济危困(王业蘧 1995)。树木死亡是森林动态变化中一项重要而复杂的生态过程。树木死亡以后不仅能够为森林生态系统中其他个体提供资源，还能够为其他树木的更新及幼苗的更新腾出空间，同时创造了林窗条件，进而促进物种共存并维持森林群落生物多样性(Franklin et al. 1987, Canham et al. 2001, Uriarte et al. 2012)。典型阔叶红松林是小兴安岭地区的地带性顶极植被，研究典型阔叶红松林树木死亡与物种共存有助于人们更深入地理解典型阔叶红松林生物多样性维持机制。

1.7 研究内容与技术路线

1.7.1 研究内容

本研究以小兴安岭典型阔叶红松林 9 hm² 大型森林动态监测样地为研究对象，主要研究内容如下：①采用广义线性混合模型及似然比检验，探讨负密度制约对树木死亡的影响；②应用广义地理加权回归模型及地理变异检验，探讨生境异质性对树木死亡的影响；③结合广义线性混合模型等手段，研究树木死亡-生长种间权衡与物种共存之间的关系；④利用负二项回归、零膨胀回归等模型手段，探索树木死亡-更新种间权衡与物种共存之间的关系；⑤根据物种共存理论对经典树木死亡模型的部分结构进行修正，并集成森林动态监测样地数据，在微尺度下对树木死亡进行动态模拟。

1.7.2 技术路线

本研究的技术路线如图 1-3 所示。

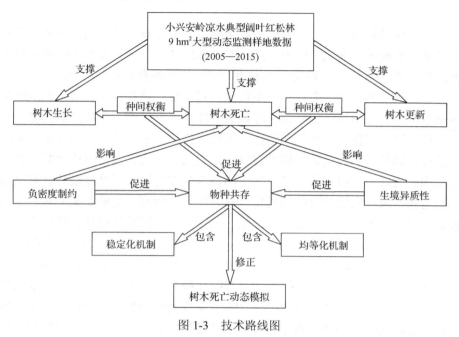

图 1-3 技术路线图

Fig. 1-3 Schematic diagram of methodologies

1.8 拟解决的关键科学问题

(1) 稳定化机制(如负密度制约)如何影响典型阔叶红松林树木死亡动态?

(2) 树木死亡-生长/更新种间权衡与物种共存理论(即稳定化机制与均等化机制)之间存在何种关系?

(3) 树木死亡动态模拟模型如何借鉴物种共存理论?

2 研究区域概况与研究方法

阔叶红松林是以红松为主的针阔混交林,主要分布在我国东北地区东部山区和俄罗斯远东地区,南北跨度为 18°(34°N～52°N),东西跨度为 16°20′(124°E～140°20′E),在朝鲜半岛,以及日本本州、四国等岛亦有分布,同时沿 50°N 线两侧也有红松树种的岛状分布。从新生代第三纪,红松就存在于我国东北地区东部山区和俄罗斯沿海边区。阔叶红松林最大限度地保存了第三纪群落的古老结构特点,红松是第三纪保留下来的孑遗种。阔叶红松林生态系统的原始状态表明,系统内的针、阔叶树种及其他生物种群已经稳定共存很久,以至形成阔叶红松林的混交结构,同时也形成在这个顶极区内的群落分布格局。阔叶红松林具有极其复杂的动植物组成,它的林分结构稳定,林分生产力较高,是欧亚大陆北温带地区物种分布最丰富的区域。在我国东北地区东部山区,阔叶红松林表现出明显的与气候水热条件有关的区域性分异。在这一范围内,形成了温度和水分条件渐变的大梯度,与此相适应,在群落特征和生物物种组成方面分异为:南部亚区(以南部阔叶红松林为代表);中部亚区(以典型阔叶红松林为代表);北部亚区(以北部阔叶红松林为代表)。在 44°30′N 以北、48°20′N 以南的小兴安岭山地,分布着典型阔叶红松林(王业蘧 1995)。典型阔叶红松林是目前我国保存下来最完整和典型的原始红松针阔混交林分布区之一,也是中国和东北亚地区极具典型代表性的温带原始林区。但是,由于人口的增长和砍伐活动的频繁,阔叶红松林生态系统现在处于资源和生态危机的阶段。为了对阔叶红松林生物多样性进行长期监测与研究,2005 年在黑龙江凉水国家级自然保护区内建立了典型阔叶红松林 9 hm^2 大型森林动态监测样地。

2.1 研究区域概况

2.1.1 地理条件

研究区位于黑龙江凉水国家级自然保护区内(47°10′50″N, 128°53′20″E)。保护区位于黑龙江省伊春市带岭区,总面积 121 339 hm^2,地处小兴安岭南部达里带岭支脉的东坡,海拔 280～707 m,山地坡度一般为 10°～15°,为典型的低山丘陵地貌。境内密被森林,森林覆盖率高达 96%,分布着大片较原始的典型阔叶红松林,是我国目前保存下来的最为典型和完整的原生阔叶红松混交林分布区之一。保护区内现有原始成过熟林面积 4100 hm^2,其中红松林面积占 80%。

2.1.2 气候概况

该地区地处中纬度大陆东岸，属于温带大陆性季风气候，四季分明。春天迟缓，降水少，多大风，易发生干旱；夏季多受副热带变性海洋气团的影响，降水集中，气温较高且短促；秋季降温急剧，常出现早霜；冬季多在变性极地大陆气团控制下，气候严寒、干燥且漫长。由于该地区的纬度较高，太阳辐射较少，所以年平均气温低，为–0.3℃左右。年平均最高气温为7.5℃左右，年平均最低气温为–6.6℃左右，>10℃积温为1700℃。年平均降水量676 mm，多集中在6~8月，占全年降水量的60%~70%。年平均相对湿度为78%，年平均土壤温度为1.2℃，无霜期100~120 d(5月中下旬至9月中下旬)，积雪期130~150 d，河流结冻期长达6个月(10月下旬至4月中下旬)。

2.1.3 土壤概况

黑龙江凉水国家级自然保护区内土壤的垂直分布不明显，只有地域性分布规律。其中，地带性土壤为暗棕壤，占自然保护区总面积的84.91%，分布于山地；非地带性土壤包括沼泽土、草甸土及泥炭土，分别占自然保护区总面积的13.07%、1.20%及0.82%，其中草甸土分布在林中空地和河流两岸的阶地上，沼泽土和泥炭土均分布于河流两岸的低洼地和山间谷地排水不良的地段。

2.1.4 植被类型

黑龙江凉水国家级自然保护区内的地带性顶极植被群落是以红松为主的温带针阔混交林，属于温带针阔叶林地北部亚地带。该区在整个阔叶红松林分布区属于典型阔叶红松林分布亚区。阔叶红松林主要建群树种为红松，伴生着温带阔叶树种，包括紫椴(*Tilia amurensis*)、糠椴(*Tilia mandshurica*)、色木槭(*Acer mono*)、青楷槭(*Acer tegmentosum*)、花楷槭(*Acer ukurunduense*)、裂叶榆(*Ulmus laciniata*)、水曲柳(*Fraxinus mandschurica*)、蒙古栎(*Quercus mongolica*)、大青杨(*Populus ussuriensis*)、黄檗(*Phellodendron amurense*)、胡桃楸(*Juglans mandshurica*)等20余种，同时也伴生着欧亚针叶林中的寒温性树种，包括臭冷杉(*Abies nephrolepis*)、红皮云杉(*Picea koraiensis*)、鱼鳞云杉(*Picea jezoensis*)等。林下灌木和藤本植物有刺五加(*Acanthopanax senticosus*)、龙牙楤木(*Aralia elata*)、瘤枝卫矛(*Euonymus pauciflorus*)、毛榛(*Corylus mandshurica*)、狗枣猕猴桃(*Actinidia kolomikta*)、五味子(*Schisandra chinensis*)、山葡萄(*Vitis amurensis*)等。草本和蕨类植物包括荨麻(*Urtica fissa*)、草地风毛菊(*Saussurea amara*)、猴腿蹄盖蕨(*Athrrium brevjrons*)、蚊子草(*Filipendula palmate*)、薹草(*Carex* spp.)、山茄子(*Brachybotrys paridiformis*)、木贼(*Equisetum* spp.)、山芹(*Ostericum* spp.)等。

2.2 数据收集

2.2.1 树木动态监测数据

于2005年在保护区建立了一块典型阔叶红松林9 hm²(300 m×300 m)大型森林动态监测样地(forest dynamics plot, FDP)(图2-1)。该样地是中国森林生物多样性监测网络(Chinese Forest Biodiversity Network, CForBio)中的一个节点。参照美国热带森林研究中心-全球森林生物多样性监测网络(Center for Tropical Forest Science-Forest Global Earth Observatory Network, CTFS-ForestGEO)技术规范(http://www.forestgeo.si.edu/),将样地分为900个10 m×10 m的标准样方。样地平均海拔约463 m。第一次调查在2005年,测定所有胸径(DBH)≥2 cm的木本植物的胸径、坐标,鉴别树种并编号挂牌;2007年使用Vertex III (Haglöf Sweden AB, Långsele, Sweden)超声波测高仪对样地内所有挂牌的活立木进行补充测量,测定树高,并对树高＞10 m的树木进行冠幅调查,即以树干为中心测量8个方向(东、西、南、北、东南、西北、东北和西南)的冠幅长度。在2010年进行第二次调查,除了对样地中所有已挂牌的活立木测量其胸径及存活状态(活或者死)之外,还对样地内2005～2010年更新的所有DBH≥1 cm的木本植物活立木进行调查,测得其胸径、坐标、鉴别树种并编号挂牌。在2015年进行第三次调查,对样地中所有之前已挂牌的活立木进行复查,测量胸径及存活状态(活或者死),并对样地内2010～2015年间更新的所有DBH≥1 cm的木本植物活立木进行调查,测得其胸径、坐标、鉴别树种并编号挂牌。

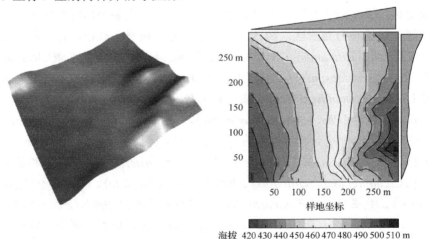

图2-1 小兴安岭凉水典型阔叶红松林9 hm²动态监测样地地形图

Fig. 2-1 Topographic map of the Liangshui 9 hm² typical mixed broadleaved-Korean pine FDP, China

2.2.2 地形和土壤状况

在 ArcGIS 10.1 软件中，根据 1 m 间隔的样地等高线矢量数据(ESRI Shapefile)分别创建 5 m、10 m 和 20 m 空间分辨率的数字高程模型(digital elevation model, DEM)栅格数据，这三个空间分辨率分别对应 5 m、10 m 和 20 m 尺度的样方大小。根据 DEM 数据可分别生成坡度(slope)、坡向(aspect)、凹凸度(convexity)、曲率(curvature)和山体阴影(hillshade)等诸多栅格数据格式的地形因子。其中，坡度和坡向采用基于栅格 Moore 邻域(即 8 邻域)的三阶反距离平方权差分算法(Burrough and McDonell 1998)。凹凸度定义为中心样方的海拔减去周围 Moore 邻域样方海拔的平均值(Valencia et al. 2004)。当凹凸度为正值时，说明相比于邻近样方，该样方整体上是向上凸起的，向外排水；当凹凸度为负值时，说明相比于邻近样方，该样方整体上是向下凹陷的，向内汇水。凹凸度的值越大，则说明中心样方越凸起；凹凸度的值越小，则说明中心样方越凹陷。凹凸度常用来对某地区的水分含量进行重新分配。凹凸度与另一常见地形因子——曲率(curvature)的具体算法虽然存在一定差异(曲率的算法基于输入表面栅格 Moore 邻域的二阶导数)，但两者均用来表征对土壤水分的重新分配。曲率的值越大，说明中心样方越向外排水；曲率的值越小，则说明中心样方越向内汇水，这两个地形因子选取其中任一项即可代替另一项(Zevenbergen and Thorne 1987, Moore et al. 1991)。例如，在 5 m 样方尺度下，凹凸度与曲率具有显著的、极强的相关性($r = 0.99$, $P < 0.0001$)。坡向和年平均太阳方位角两者共同决定了样方是阴坡还是阳坡。例如，本研究区的阴坡与阳坡通常参考了该地区的年平均太阳方位角进行如下划分：阴坡(337.5°～22.5°, 22.5°～67.5°)，半阴坡(67.5°～112.5°, 292.5°～337.5°)，平地，半阳坡(112.5°～157.5°, 247.5°～292.5°)，阳坡(157.5°～247.5°)，共 5 个等级。此外，由于坡向用 0°～360°来表示，在分析、建模应用时不能直接使用，常常需要进行正弦、余弦函数等数学转换，将其转换为两个因子。当坡度小于年平均太阳高度角时，坡度和年平均太阳高度角两者共同影响阴坡的相对光照强度。以往的研究中多采用坡度和坡向两项地形因子，但是几乎未见关于应用山体阴影的报道。山体阴影综合考虑了坡度、坡向、年平均太阳方位角和高度角(或天顶角)，可以表征某个特定位置(如某样方)的相对太阳辐射强度，具体算法可参见文献(Burrough and McDonell 1998)。研究区域年平均太阳方位角和高度角分别约为 202.5°和 45°。山体阴影的值越大，说明相对光照强度越大；山体阴影的值越小，说明相对光照强度越小。综上所述，本研究在充分考虑了地形因子作为间接生态因子所具有的对太阳辐射、土壤水分重新分配作用的前提下，选用海拔、凹凸度和山体阴影作为地形因子。此外，按照 5 m 样方尺度对坡位进行了调查，

以获取树种邻体半径。

为了捕获微尺度下的土壤属性空间变异,将土壤样品带回实验室测定土壤理化性质,共 10 项,具体包括土壤全磷、土壤有效磷、土壤全氮、土壤水解氮、土壤有机碳、土壤速效钾、土壤 pH、土壤容重、土壤体积含水率和质量含水率。2013 年 8 月,将典型阔叶红松林 9 hm^2 森林动态监测样地划分为 225 个 20 m×20 m 的大样方。以 256 个大样方交叉点作为采样基点,从每个基点的北、东北、东、东南、南、西南、西或西北 8 个方向中随机选取任一方向,在选定的方向上距离基点 2 m、5 m 或 8 m 处随机选取两处作为采样点(Webster and Oliver 2007),共计采集 768 个土壤样品,采样深度为 0~10 cm。采样时,清除采样点表层枯落物的腐殖质层,在该点周围 0.5 m 范围内用直径为 5 cm 的土钻钻取 3 个土样,将 3 个土样充分混合来代表该采样点的土壤样本。采集好的土样放入密封袋内带回实验室,仔细去除样品中根系和石砾等杂质,取出一部分新鲜土用烘干法进行土壤质量含水率测定,土壤质量含水率为土壤中所含水分重量占烘干土重量的百分数。余下的土壤自然风干后用于土壤理化性质的测定。取一部分风干土样过 2 mm 土壤筛,用 1 mol/L 的盐酸处理 16~24 h 后,在烘箱内 105℃烘干 3 h 去除无机碳,利用 multi N/C 2100 分析仪(Analytik Jena AG, Jena, Germany)测定土壤有机碳。土壤 pH 采用电位法测定,土与水溶液之比为 1∶2.5。土壤全氮使用 Hanon K9840 凯氏定氮仪(Jinan Hanon Instrument Co., Ltd., Jinan, China)测定。土壤水解氮采用碱解-扩散法测定。土壤体积含水率采用时域反射计(time domain reflectometry, TDR)测得(IMKO, Ettlingen, Germany)。土壤速效钾采用乙酸铵-火焰光度计法测定。土壤有效磷采用 0.05 mol/L 盐酸-0.025 mol/L 硫酸浸提法测定。采用环刀(100 ml)采集土样并进行土壤容重的测定。土壤全磷采用硫酸-高氯酸酸溶-钼锑抗比色法测定。

2.2.3 系统发育树和功能性状数据

本研究借助 Phylomatic(v3)在线软件(http://phylodiversity.net/phylomatic/)构建典型阔叶红松林群落的系统发育树,生成基于被子植物种系发生学组(angiosperm phylogeny group, APG)III 分类系统具有进化枝长的系统发育树框架(Bremer et al. 2009, Zanne et al. 2014)。由于 Wikstroem 的分化时间已经发表了十几年,在此期间,被子植物各类群的分化时间多有变化,因此通过 Phylocom BLADJ 的方法校正分化时间已经有些过时。从 2015 年开始,Phylomatic 网站整合了 Zanne 等(2014)的进化树骨架,可以基于该进化树直接获得有枝长的进化树,而无须 Phylocom BLADJ 模块及 Wikstroem 的时间文件(Webb and Donoghue 2005)。通过 Phylomatic 在线软件,可以直接获得有枝长的进化树。

以 320 株样地周边的树木作为样本,研究测量了木本植物的比叶面积(specific leaf area, SLA)、最大树高(maximum tree height, H_{max})及木质密度(wood density, WD)共 3 项植物功能性状指标。比叶面积是植物对叶片构建投入-回报的权衡,最大树高是接受太阳辐射持续性与繁殖时间之间的权衡,木质密度的差异则反映水力安全-有效性的权衡,这 3 项功能性状指标与树木死亡及生长关系密切。Díaz 等(2016)的研究结果显示,在全球尺度上,这 3 个性状代表了物种、叶经济谱和植物大小之间生活史变化的两个主轴上的大部分功能性状变异。尽管如此,在本研究中,这 3 个性状仍然具有较高的相关性,因此采用主成分分析(principal component analysis, PCA)将其浓缩成 2 个主成分轴,去除相关性。叶片性状采集方法参照 Pérez-Harguindeguy 等(2013)的手册。每个物种尽量选取 5 个健康成年个体,每个个体选取 20 片充分暴露于阳光下的叶片。比叶面积为叶面积与叶片烘干重(65℃)的比值。物种的最大树高依据《中国植物志》。木质密度为烘干后(103℃, 72 h)的质量除以新鲜状态下的体积,体积采用排水法测定。在距离森林动态样地边界处较近的地点,每个树种各采集 3~5 株。对于大树(DBH>10 cm),其采集的生长芯被分成 1 cm 的片段,木质密度用树木横截面环形加权平均值(annulus weighed)来计算。对于灌木,主干直接砍断后测定木质密度。本研究共采集了典型阔叶红松林 40 个主要木本植物的上述功能性状数据。

2.3 主要分析方法

本研究以模型为主要手段,辅以常见的统计分析。运用模型手段解决生态学问题是现代生态学领域重要的研究方法。生态学领域运用的模型一般可分为两类:经验模型(empirical model)和过程模型(process-based model)。前者亦称为统计模型,后者可称为机理模型。经验模型一般是应用统计学理论,根据系统变量之间回归关系所建立的模型。经验模型的优点是具有较好的精确性(precision)和真实性(reality),但是模型概括性(generality)不强,应用范围较窄,并且需要大量的树木生长数据,模型参数一般不具有明确的生物学意义,若建立动态模型,则时间步长(时间分辨率)依赖于具体的野外动态监测时间间隔。机理模型一般涉及较多的植物生理生态学过程和机理,多为计算机仿真模型。机理模型具有较强的概括性和真实性,模型参数通常均具有明确的生物学意义,但是精确性不如经验模型(Levins 1966, Sharpe 1990)。两类模型各有其擅长之处,并无优劣之分,使用得当方可充分发挥其作用。基于模型内部属性的模型分类认为,统计模型、过程模型和理论模型在模型真实性、精确性及概括性等方面各有权衡和侧重(图 2-2)。

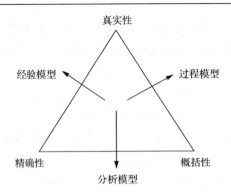

图 2-2 基于模型内部属性的模型分类(Levins 1966, Sharpe 1990)
Fig. 2-2 A classification of models based on their intrinsic properties(Levins 1966, Sharpe 1990)

2.3.1 统计模型

统计模型,即经验模型,又被称为基于调查数据的建模方式(inventory-based modeling approach)。根据关注点的不同,统计模型仍可进一步细分为预测模型(predictive model)和解释模型(descriptive model)两类(Adler 2012)。其中,预测模型主要关注模型的预测精度,模型自变量及其参数估计值对因变量的贡献(包括大小、正负和显著性)方面的讨论处于次要地位。通常以各类模型精度评价指标(如受试者操作特性曲线,receiver operating characteristic curve, ROC 曲线)评判模型预测精度。以对树木死亡建模为例,预测模型多采用广义线性模型中的 Logistic 回归分析,自变量多为单木大小、单木竞争、单木活力、林分密度、立地条件等(Hamilton 1986)。这一类以精确预测树木死亡为主要目的的模型,多应用于森林经营管理。该类预测模型在林分生长与收获模型分类中常被称为单木枯损模型。解释模型主要关注模型自变量及其参数估计值对因变量的贡献(包括大小、正负和显著性)。但是,模型精度在解释模型当中处于次要地位,甚至在模型自变量当中还包含着一些对因变量贡献很小甚至几乎没有显著影响的自变量,而在解释模型当中设置这种对因变量贡献甚微的自变量对于验证某些假说和理论是有价值的。以对树木死亡建模为例,解释模型也同样采用广义线性模型中的 Logistic 回归分析,但是自变量的设置通常对应某些假说和机理,一般设置为对树木死亡具有潜在影响的生物与非生物等生态因子。生物因子主要包括生物邻体变量,通常对应生物互作(负密度制约假说)方面的检验(Zhu et al. 2015b)。非生物因子主要包括地形和土壤等生境变量,通常对应树种生境偏好等方面的检验(Wang et al. 2012)。解释模型很少进行模型精度方面的评价。各类统计模型的参数估计通常可由统计软件计算完成。近年来,R 软件以其免费、开源、扩展性好等诸多优点越来越受到科研人员,尤其是生态学者的青睐,逐渐成为各类生态学分析的主流统计分析软件，是统计计算与制图的强大工具。

本研究在第 3 章、第 4 章、第 5 章和第 6 章均应用了统计模型当中的解释模型，包括广义线性混合模型及其子模型(随机截距模型和随机系数模型；Logistic 线性混合模型和一般线性混合模型)，以及相应的似然比检验、广义地理加权回归模型及其子模型(一般地理加权回归模型和半参数化地理加权回归模型)、相应的地理变异检验、负二项回归模型、零膨胀回归模型、零膨胀负二项回归模型。各类模型的特点参见各章研究方法部分(见 3.2 节、4.2 节、5.2 节、6.2 节)。

2.3.2 过程模型

过程模型，即机理模型，通常又被称为基于过程的建模方式(process-based modelling approach)。机理模型的建立方式与以统计学理论为基础的经验模型具有本质差别。机理模型的构建通常以植物生理生态学过程和机理为理论基础，辅以面向对象的程序设计(object-oriented programming, OOP)等软件工程概念与技术进行模拟实现。机理模型当中的输入(input)对应统计模型当中的自变量，机理模型当中的输出(output)对应统计模型当中的因变量。统计模型通常将自变量预测因变量的中间过程视为黑箱，即对内部结构和特性不清楚。而与此相反，机理模型的主要特征是对中间过程的细致刻画，中间过程往往视为白箱或者灰箱，即对内部结构和特性比较清楚或一般清楚。以对树木死亡建模为例，模型的输入通常与树木死亡的经验模型无太大差别。模型的结构却以树木生理生态学为核心，树木的死亡被定义为内禀死亡和外部胁迫死亡两者的综合作用(Bugmann 2001)。内禀死亡受到树种年龄和树木径级影响，外部胁迫死亡受到邻体竞争、温度、湿度及树木生长效率影响，而树木的生长又与树冠垂直方向上平均光合作用与树干边材呼吸损耗量密切相关。机理模型的参数化是除了模型结构搭建之外的另一个重要过程。与经验模型依托数理统计学进行模型参数估计不同，机理模型的参数通常需要采用野外实测的方式获取，参数具有明确的生物学意义。此外，基于过程的机理模型的时间步长不依赖于具体的野外动态监测时间间隔。C#是一种由 C/C++衍生出的功能强大的、通用的、面向对象的计算机编程语言，已成为编写机理模型程序的主流编程语言之一。另外，在.NET Framework 框架下，C#与 R 可实现混合编程，这类混编方式充分发挥了两类语言的各自优势，可为构建半统计半机理模型提供技术支撑。

本书第 7 章在参考以往大量经典森林动态模型的基础上，经过改进后研制了树木死亡过程模型。树木死亡过程模型详情参见第 7 章研究方法部分(7.2 节)。

3 负密度制约对树木死亡的影响

3.1 引　　言

　　生物互作通常可以指代负密度制约效应，它是森林群落物种共存的重要机制之一(HilleRisLambers et al. 2012)，同时也是影响树木存活的一项重要的生物因子。通常，生物邻体对树木存活的效应存在种间变异，同种邻体对树木存活效应的种间变异的变异幅度比异种邻体更广(Comita et al. 2010, Johnson et al. 2012)。根据巴拿马 BCI 热带雨林 50 hm² 大型森林动态监测样地的研究发现，同种邻体指数效应的种间变异幅度高于异种邻体谱系相异性指数(Zhu et al. 2015b)。各个树种由于生活史策略存在差异，导致生物与非生物因子对其影响也存在不同。有研究发现，物种叶习性(leaf habit)可能是决定物种对同种邻体密度敏感性的主要影响因素(Comita et al. 2010, Bai et al. 2012, Lu et al. 2015)。

　　邻体影响半径可能会影响生物邻体对树木存活效应种间变异的侦测(detection)。以往的研究通常采用统一、固定的邻体半径来计算生物邻体因子，忽略了邻体半径存在种间变异的事实。虽然诸多关于森林中树木竞争效应(Biging and Dobbertin 1992)及森林动态模型研制(Smith and Urban 1988, Busing 1991, Canham et al. 2004)等的研究均采用统一、固定的邻体影响半径，但是这有待商榷。Zhang 等(2010)发现假色槭(*Acer pseudosieboldianum*)和白牛槭(*Acer mandshuricum*)的空间分布在 10 m 范围内均存在显著的负相关，然而，假色槭和簇毛槭(*Acer barbinerve*)的空间分布仅在 2 m 范围内存在显著相关关系。Lavoie 等(2007)发现在黑云杉(*Picea mariana*)林采伐迹地上，树木累计生长量与土壤有机碳层厚度在 20 m 以内存在显著负相关，然而在黑云杉林火烧更新迹地上则不存在这种趋势。由此可见，不同树种存在不同的空间关系，邻体半径大小可能会影响树种如何对生物邻体效应做出响应。对所有树木均采用统一的、固定的邻体半径来构建生物因子，可能会导致无法侦测到受空间尺度影响的密度制约效应的真实检测结果，并且妨碍不同研究之间关于密度制约效应大小的比较(Wang et al. 2012, Comita et al. 2014, Zhu et al. 2015b)。然而到目前为止，有关邻体影响半径存在种间变异并且在温带森林开展的密度制约研究鲜有报道。

　　在本章中，使用 2005～2015 年共 10 年间的树木死亡动态监测数据，探究负密度制约如何影响典型阔叶红松林树木死亡动态，并试图回答生物邻体因子对树木存活的效应是否存在显著的种间变异。具体分为以下 3 个小问题：①生物邻体因子对树木存活的效应是否存在显著的种间变异？②邻体影响半径是否会对这种变异的检测造成影响？③生活史策略可否对上述变异做出合理解释？

相对应地提出 3 点假设：①生物邻体因子对树木存活的效应存在显著的种间变异；②邻体影响半径能够对这种变异的检测结果造成影响；③树种叶习性间的差异可能是上述变异的原因之一。

3.2 研 究 方 法

3.2.1 数据收集

不同于树木生长和更新动态，树木死亡是一个相对缓慢的过程。为了尽可能研究较长时间段内的树木死亡动态，本章选取 2005～2015 年共 10 年间的树木死亡动态监测数据集。2005 年，对样地内所有的 DBH≥2 cm 的活的木本植物进行编号挂牌、测量胸径、定位坐标并鉴别树种。在 2015 年，对 2005 年调查的活立木进行复查，记录其存活状态(存活或死亡)。地形与土壤数据收集参见 2.2.2 节。

3.2.2 个体尺度树木死亡驱动因子的构建

本章在树木个体尺度上构建两个生物邻体因子，即同种邻体指数(conspecific neighbor index, CI)和异种邻体平均谱系相异性指数(average phylogenetic dissimilarity index of heterospecific neighbors, PI)。这两项生物邻体因子的计算公式参考了 Chen 等(2016)提出的生物邻体指标计算公式。邻体树木尺寸(如胸高断面积)越大(Comita and Engelbrecht 2009)，并且与基株的空间距离越近，则邻体对基株的生长和死亡影响也越大(Uriarte et al. 2010)。因此，在计算异种邻体平均谱系相异性指数时，本章充分考虑了邻体树木尺寸和空间距离两项因素。对空间距离权重的处理，本章使用了高斯核函数(Gaussian kernel function)(Fotheringham et al. 2003, Pu et al. 2017)。高斯核单调递减权重函数不仅考虑了空间距离，还考虑了邻体半径种间变异，而以往只考虑了空间距离，因此，采用高斯核作为权重函数优于以往的反距离加权函数。根据邻体组成、初始调查时测量的胸高断面积、基株所在样方的坡位、系统发育树及高斯核空间距离权重函数，在 3 种邻体半径条件下(固定半径 5 m、固定半径 10 m 及考虑种间变异条件下的邻体半径)，分别计算上述两个生物邻体因子，并选取最优邻体半径进行分析。同种邻体指数(CI)和异种邻体平均谱系相异性指数(PI)不仅考虑了邻体数量，还考虑了空间距离权重、邻体树木尺寸及系统发育距离，是具有复杂单位的、综合性的生物因子。

杨光(2006)认为邻体影响半径应由基株所属树种及其生长地点的坡位共同决

定，根据生长方程回归分析法和生长释放法估测得到了东北地区东部山区主要树种邻体影响半径(表 3-1)。根据此研究发现，亲缘关系越近的树种可能具有相似的邻体半径(例如，白桦 *Betula platyphylla* 和枫桦 *Betula costata*)。据此，本章假设树种邻体半径在系统发育结构上具有一定的保守性，故借助于系统发育树和种间系统发育距离(Ma)，近似估计其他未知树种的邻体半径。

表 3-1 东北东部山区主要树种邻体影响半径(杨光 2006)   (单位: m)
Table 3-1 Neighbour influence radii (m) of some major species in the eastern mountains of Northeast China

树种	坡位			测定方法
	山脊	山坡	山谷	
臭冷杉 *Abies nephrolepis*	5	7	6	生长方程回归分析法(杨光 2006)
枫桦 *Betula costata*	6	5	5	同上
白桦 *Betula platyphylla*	6	5	5	同上
红皮云杉 *Picea koraiensis*	7	6	6	同上
红松 *Pinus koraiensis*	6	5	6	同上
胡桃楸 *Juglans mandshurica*	5	5	5	同上
水曲柳 *Fraxinus mandshurica*	6	6	7	生长释放法(杨光 2006)

生物邻体因子公式定义如下:

$$\mathrm{CI} = \sum\nolimits_{i}(\mathrm{CBA}_i \cdot W_i) \tag{3-1}$$

$$\mathrm{PI} = \sum\nolimits_{i}(\mathrm{PD}_i \cdot \mathrm{HBA}_i \cdot W_i) / N \tag{3-2}$$

$$W_i = \mathrm{Exp}\left[-\frac{1}{2}\left(\frac{\mathrm{SD}_i}{R}\right)^2\right] \tag{3-3}$$

式中，CI 代表同种邻体指数；PI 代表异种邻体平均谱系相异性指数；CBA_i 为同种邻体胸高断面积(m^2)；HBA_i 为异种邻体胸高断面积(m^2)；i 代表树牌号；W_i 代表高斯核空间距离权重函数；PD_i 代表基株所属树种与邻体所属树种之间的系统发育距离(Ma)；SD_i 代表基株与邻体之间的欧氏空间距离(m)；R 代表邻体影响半径(m)；N 代表异种邻体株数。

近年来，群落系统发育学被引入群落生态学并为此类研究提供了新的视角。群落系统发育学的基本假设之一是物种生态位保守假说，即物种在进化过程中生态位是保守的。换言之，亲缘关系越相近的物种其生态位越相似，因此异种邻体间的资源竞争也更加强烈。简单而粗略地把邻体半径内的邻体物种分成同种和异种两类，会模糊不同物种对基株物种影响的巨大差异性(Pacala et al. 1996)。为了研究系统发育对树木死亡的影响，本章构建了异种邻体平均谱系相异性指数代替以往研究中大量使用的异种邻体胸高断面积指数，后者未考虑群落生态学理论的最新发展。早期的密度制约研究只考虑了邻域内邻体数量对基株的影响，近年来，随着密度制约研究的进一步发展，邻体大小、基株与邻体间空间距离及系统发育距离等因素也逐渐应用到了密度制约指数的构建当中。

以往研究认为密度制约研究中必须考虑生境异质性的影响(Piao et al. 2013)。地形因子包括海拔、凹凸度和山体阴影。由于土壤因子之间存在较强的相关性，因此采用主成分分析对10项土壤因子进行降维及去除相关性处理，提取了3个主成分轴代替原土壤指标。

3.2.3 个体尺度树木死亡模型

本章借助于统计学中的广义线性混合效应模型(generalized linear mixed-effects model, GLMM)及似然比检验(likelihood ratio test, LRT)，试图解答生物邻体因子对树木存活的效应是否存在显著的种间变异(Bolker et al. 2009)。

使用广义线性混合模型建立个体尺度上的树木存活模型，因变量为树木在2015年的存活状态，即1代表存活，0代表死亡。因变量服从二项分布(binomial distribution)，模型采用逻辑连接函数(Logistic link function)(Bolker et al. 2009)。以往研究发现，树木初始大小能够显著地影响树木存活，因此将自然对数转换后的初始胸径作为模型的固定效应(fixed effect)之一(Uriarte et al. 2004)。固定效应还包括生物(生物邻体因子)与非生物因子(地形和经过 PCA 处理后的土壤因子主成分)。该树木存活模型中包含了两项交叉随机效应(cross random effect)：第一个随机效应是树种名称，这是由于不同树种的基准(baseline)存活率和生长率存在差异；第二个随机效应是20 m 尺度的大样方编号，这是由于同一个大样方内的树木通常具有相似的存活率和生长率(空间自相关性)(Zhu et al. 2015b, Zhang et al. 2016b)。

利用不同类型的变量组合构建8个候选模型(candidate model)，分别为零模型(仅包含初始胸径变量)、密度制约模型(在零模型基础上添加生物邻体变量)、地形模型(在零模型基础上添加地形变量)、土壤模型(在零模型基础上添加 PCA 后的土壤主成分变量)、地形与土壤模型(在零模型基础上添加地形与 PCA 后的土壤主成分变量)、密度与地形模型(在零模型基础上添加生物邻体与地形变量)、密度与土壤模型(在零模型基础上添加生物邻体与 PCA 后的土壤主成分变量)及全模型(包含所有变量)。根据赤池信息准则(Akaike's information criterion, AIC)判定最简约模型。采用标准差标准化法(Z score)对所有自变量进行无量纲化处理，即所有样本各自变量减去各自自变量的平均值后再除以标准差,以便允许各个自变量参数之间可以直接地比较大小(Gelman and Hill 2006)。此外,为了排除边缘效应干扰,在建模时边缘样方中的树木不作为基株,但仍可作为邻体树木。

为了判定生物邻体对树木存活的效应是否存在显著的种间变异,本研究通过比较广义线性混合模型中的两个子模型——随机截距模型(random intercept model)和随机系数模型(random coefficient model)的方式来判定某一生物邻体变量对树木存活的效应是否存在差异,并借助似然比检验来判定这种差异在统计学上是否显著。这里所讲的"效应(effect)"是指模型中某个自变量对因变量的贡献,换言之,它是指模型中的自变量参数估计值(parameter estimate)。随机截距模型和随机系数模型的唯一区别是:前者某个自变量的参数(斜率)是固定不变的,而后者该自变量的参数(斜率)是变化的,后者既包含随机截距,也包含随机斜率(random slope),两者统称为随机系数。为了保证有足够的统计功效(statistical power),本章选择多度不少于 40 的树种(共计 22 个树种)进行此分析。

为了判定邻体半径能否影响生物邻体对树木存活效应种间变异的侦测,本章采用变化的邻体半径(包括根据以往研究得出的东北东部山区主要树种邻体半径和根据亲缘关系近似估测的其他树种的邻体半径),以及统一的、固定的邻体半径(5 m 和 10 m)分别建立树木存活模型,并通过比较随机截距模型、随机系数模型及借助似然比检验,分别判定在 3 种不同邻体半径的条件下,随机截距模型和随机系数模型是否存在显著差异。

本章采用的广义线性混合模型构建如下:

$$\text{Logistic}(p_{ij}) = \ln\left(\frac{p_{ij}}{1-p_{ij}}\right) = \sum_k (\beta_k \cdot x_{ij}) + \Phi_h \tag{3-4}$$

式中,p_{ij} 代表树种 j 个体 i 的模型预测存活概率;x_{ij} 代表树种 j 个体 i 的固定效应取值;Φ_h 代表解释因变量种间变异和空间自相关性的交叉随机效应;β_k 代表固定效应参数估计值。

广义线性混合模型的两个亚模型——随机截距模型与随机系数模型构建如下(以同种邻体指数 CI 的参数估计值为例)。

随机截距模型:

$$\text{Logistic}(p_{ij}) = \beta_{0j} + \beta_1 \cdot \text{DBH}_{ij} + \beta_2 \cdot \text{CI}_{ij} + \beta_3 \cdot \text{PI}_{ij} + \beta_4 \cdot \text{Elevation}_{ij} \\ + \beta_5 \cdot \text{Convexity}_{ij} + \beta_6 \cdot \text{Hillshade}_{ij} + \Phi_h \tag{3-5}$$

随机系数模型:

$$\mathrm{Logistic}(p_{ij}) = \beta_{0j} + \beta_1 \cdot \mathrm{DBH}_{ij} + \beta_{2j} \cdot \mathrm{CI}_{ij} + \beta_3 \cdot \mathrm{PI}_{ij} + \beta_4 \cdot \mathrm{Elevation}_{ij} \\ + \beta_5 \cdot \mathrm{Convexity}_{ij} + \beta_6 \cdot \mathrm{Hillshade}_{ij} + \varPhi_h \tag{3-6}$$

同资源种团/功能群(guild/functional group)分析：为了解释生物邻体因子对树木存活的效应存在种间变异，本章根据树种生活史策略及其资源利用方式将树种分为不同的功能群，即同资源种团。根据叶习性将树木划分为落叶树种和常绿树种两个同资源种团，并且分别为每一个同资源种团建立树木存活模型，得到各个模型参数估计值的大小、正负和显著性，通过分析比较不同同资源种团模型之间的差异，探讨生活史策略对上述种间变异的影响。

广义线性混合模型和似然比检验过程采用 R 3.1.3(Team 2015)软件中的"lme4"(Bates et al. 2015)程序包实现。广义线性混合模型的参数估计采用极大似然估计法(maximum likelihood estimation, MLE)(Bolker et al. 2009)。

3.3 结　　果

3.3.1 最优变量组合

根据 AIC 模型拟合优度判别准则，在 8 个候选模型中得出密度+地形模型是最优模型，初始胸径、生物邻体和地形因子为最优变量组合(表 3-2)。此外，密度+地形模型与地形+土壤模型、密度+土壤模型、全模型没有显著差异。因此，在个体尺度分析中，本章采用初始胸径、生物邻体变量和地形因子作为自变量，重点关注生物邻体变量和地形因子对树木存活的影响。

表 3-2　基本模型选择
Table 3-2　Basic model selection

候选模型	赤池信息准则	χ^2	P
零模型	10 054	44.742	<0.000 1
密度制约模型	10 048	34.987	<0.000 1
地形模型	10 023	7.215 4	0.017 2
土壤模型	10 049	33.172	<0.000 1
地形土壤模型	10 028	0	1
密度地形模型	**10 019**	—	—
密度土壤模型	10 044	0	1
全模型	10 025	0.285 9	0.729 8

注：最优模型的 AIC 值显示为粗体。

3.3.2 个体尺度树木死亡的生物邻体驱动因子

通过比较不同邻体半径下的模型发现,当采用变化的邻体半径(树种真实或根据亲缘关系近似得到的邻体半径)下建立的模型具有最小的 AIC 值时,说明该模型是拟合优度最佳的最简约模型,即最优模型(表 3-3)。

表 3-3 不同邻体半径条件下所建模型的赤池信息准则
Table 3-3 Model selection via AIC at different neighbourhood radii

邻体半径	赤池信息准则
变化半径	**10 019**
固定 5 m 半径	10 026
固定 10 m 半径	10 026

注:最优模型的 AIC 值显示为粗体。

除了当采用固定 5 m 半径条件下建立的模型之外,本章结果发现,在固定 10 m 和变化的邻体半径条件下建立的模型中,同种邻体指数(CI)和异种邻体平均谱系相异性指数(PI)对树木存活的效应存在显著($P<0.05$)或边缘显著($0.05 \leqslant P<0.10$)的种间变异(表 3-4)。

表 3-4 不同邻体半径条件下广义线性混合模型的似然比检验
Table 3-4 LRT of GLMMs at different neighbourhood radii

邻体半径	模型(随机参数)	χ^2	P
变化的半径	随机截距		
	随机系数(同种邻体指数)	4.349	0.075[#]
	随机系数(异种邻体平均谱系相异性指数)	4.205	0.081[#]
固定 5 m 半径	随机截距		
	随机系数(同种邻体指数)	3.000	0.153[NS]
	随机系数(异种邻体平均谱系相异性指数)	9.397	0.006[**]
固定 10 m 半径	随机截距		
	随机系数(同种邻体指数)	3.944	0.093[#]
	随机系数(异种邻体平均谱系相异性指数)	19.358	<0.001[***]

*** $P<0.001$; ** $0.001 \leqslant P<0.01$; * $0.01 \leqslant P<0.05$; # $0.05 \leqslant P<0.1$; NS, 不显著。

此外,落叶树种具有显著的同种邻体负密度制约效应和边缘显著的异种邻体系统发育负密度制约效应,然而常绿树种则不然(图 3-1)。换言之,叶习性是引起生物邻体效应种间变异的原因。在群落水平上，异种邻体平均谱系相异性指数对树木存活存在显著的正效应(图 3-2)。

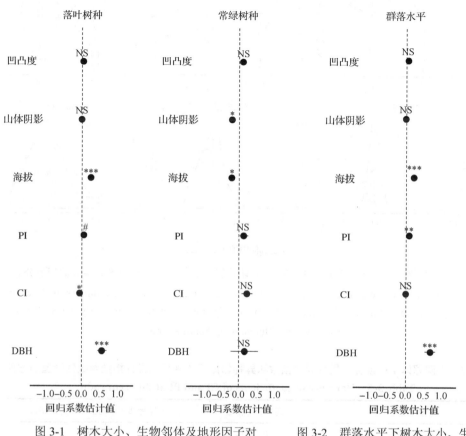

图 3-1　树木大小、生物邻体及地形因子对落叶树种与常绿树种树木存活的影响

Fig. 3-1　Standardized parameter estimates (±SE) of the effects of tree size, biotic neighbourhood and topographic variables on tree survival of deciduous and evergreen functional groups

图 3-2　群落水平下树木大小、生物邻体及地形因子对树木存活的影响

Fig. 3-2　Standardized parameter estimates (±SE) of the effects of tree size, biotic neighbourhood and topographic variables on tree survival at the community level

3.3.3　不同邻体半径下生物邻体效应种间变异检验

综合参数估计值分布与方差可得：无论采用何种邻体半径(固定 5 m、10 m、变化的邻体半径)，同种邻体指数(CI)对树木存活效应的种间变异幅度均大于异种邻体平均谱系相异性指数(PI)对树木存活效应的种间变异幅度(图 3-3、表 3-5)。

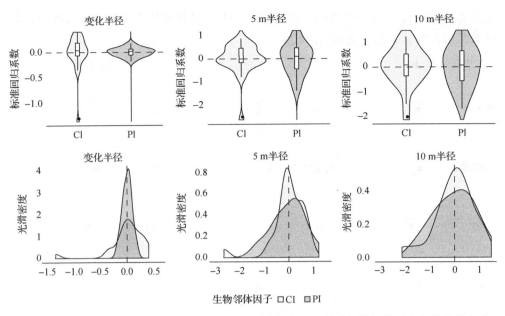

图 3-3 不同邻体半径条件下同种邻体指数和异种邻体平均谱系相异性指数参数估计值分布
Fig. 3-3 Violin plot and smooth density plot showing standardized parameter estimates of effects of the conspecific neighbour density index (CI) and average phylogenetic dissimilarity index (PI) on tree survival at different neighbourhood radii

表 3-5 不同邻体半径条件下同种邻体指数和异种邻体平均谱系相异性指数参数估计值的方差
Table 3-5　Variance of coefficients of CI and PI at different radii

邻体半径	同种邻体指数	异种邻体平均谱系相异性指数
变化半径	0.1437	0.0099
固定 5 m 半径	0.7971	0.5233
固定 10 m 半径	1.0661	0.8167

3.4 讨　论

3.4.1 生物邻体效应种间变异的影响因素

越来越多的相关研究达成了一个共识：生物邻体因子对不同物种的生长与存活的影响存在差异(Lorimer 1983, Canham et al. 2004, Comita et al. 2014)。在凉水典型阔叶红松林，生物邻体变量对树木存活的效应存在显著的或边缘显著的种间变异(表 3-4)，这符合本章的研究假设。此外，叶习性可能是引起生物邻体效应种间变异的原因(图 3-1)，这也与研究假设相符合。上述研究结果与八大公山亚热带森林动态监测样地的研究结果相似，后者研究发现，同种邻体密度或异种邻体平均谱系相异性指数对幼苗存活的影响受树种叶习性的影响(Lu et al. 2015)。落叶树种将更多的资源分配给与生长相关的功能性状，从而对植物防卫方面的投入相对减少，使得幼苗更容易受到天敌攻击而变得脆弱并导致死亡(Villar et al. 2006)；相反，常绿树种具有较强的防卫机制(如木质素和鞣酸)，因而其幼苗具有相对较高的存活概率(Coley 1988)。此外，被子植物(如本章研究的落叶树种)比裸子植物(如本章研究的常绿树种)每单位生物量需求的资源更多，从而引发更加强烈的负密度制约效应。在本研究中相同的邻体范围内，被子植物可能在与裸子植物的竞争中胜出。当裸子植物基株的邻体同样是裸子植物时，基株会有更好的生长环境和存活状态。再者，寿命较长的叶片(如针叶)对周围环境(如生物邻体)变化的敏感性相对较低。因此，综合以上分析，常绿树种树木没有受到显著的负密度制约效应影响。

本章研究结果同时揭示无论采用何种邻体半径(固定 5 m、10 m、变化的邻体半径)的条件下，同种邻体因子对树木存活效应的种间变异幅度均大于异种邻体平均谱系相异性指数(图 3-3、表 3-5)。这与其他温带和热带森林关于同种、异种和系统发育密度制约效应的研究结果类似(Comita et al. 2010, Johnson et al. 2012, Lin et al. 2012, Zhu et al. 2015b)，如巴拿马 BCI 热带雨林样地及西双版纳热带雨林动态监测样地。

再者，地形因子通常被视为独立生态地理变量以计算生态位参数，如生态位因子分析(ecological niche factor analysis, ENFA)(Hirzel et al. 2002, Basille et al. 2008)。地形因子作为间接生态因子，对土壤水分、养分空间分布可进行重新分配。本研究区海拔与土壤第二主成分轴之间也存在较强且显著的相关性($r=0.48$, $P<0.001$)。徐丽娜和金光泽(2012)研究发现本研究区(凉水)典型阔叶红松林树种分布与地形紧密关联。地形因子某种程度上，可以看成代替土壤因子的间接生态因子。这可以在某种程度上解释为什么模型最优变量组合包含地形因子，却不包含土壤因子。

3.4.2 邻体半径对检验生物邻体效应种间变异的影响

邻体半径可能影响生物邻体对树木存活效应的种间变异的检测，这与研究假设相符合。本章结果表明，当所有树木无差别地采用统一的、固定的 5 m 邻体半径时，未能检测出同种邻体效应存在显著的种间变异($P=0.153$)(表 3-4)。

关于单木竞争指数和森林动态模型中如何设置邻体半径的问题，长期以来生态学家们未能达成共识。Lorimer(1983)认为竞争指数中的搜索半径应该视基株冠幅而定。但是，在林分过程研究(Biging and Dobbertin 1992)及森林动态模型模拟研制(Smith and Urban 1988, Busing 1991, Canham et al. 2004)中，大多数关于竞争效应的研究均使用了统一、固定的邻体半径(如 10 m 或 15 m)。然而，Zhang 等(2010)在长白山温带森林发现假色槭和白牛槭在 10 m 范围内均存在显著的负相关，但是假色槭和簇毛槭仅在 2 m 范围内存在显著相关关系，超过 2 m 则无此关联性。再者，Lavoie 等(2007)发现在黑云杉林采伐迹地上，树木累计生长量与土壤有机碳层厚度在 20 m 以内存在负相关，然而在黑云杉林火烧更新迹地上则不存在这种趋势。因此，根据以上文献回顾，不难理解不同树种存在着不同的空间关系，邻体半径大小会影响树种如何对生物邻体效应做出响应。本章采用变化的邻体半径的模型是最优模型(根据 AIC 拟合优度判定标准)，这与预期相符。换言之，不能忽略邻体半径种间变异对研究结果的影响。长久以来，学者们计算同种、异种及系统发育密度制约效应的研究多数使用了统一、固定的邻体半径，这些研究很少关注邻体半径存在种间变异的事实(Webster and Oliver 2007, Wang et al. 2012, Piao et al. 2013, Lin et al. 2014, Lu et al. 2015, Wu et al. 2016)。但是，邻体半径的设置不同，可能会造成密度制约效应大小或显著性的差异，从而影响不同研究之间或同一研究内密度制约效应的比较(Comita et al. 2014, Zhu et al. 2015b)。如果未能考虑邻体半径的种间变异或邻体半径随环境的变化，那么这种疏忽可能会对密度制约研究造成不良干扰。

3.4.3 负密度制约导致的树木死亡

在群落水平树木存活分析中，异种邻体谱系相异性指数对树木存活存在显著的正效应(图 3-2)，即基株与邻体之间的亲缘关系越远，则基株的存活率越高，反之亦然，这说明群落整体上存在显著的系统发育负密度制约死亡效应。基株与邻体之间的亲缘关系越远，则两者之间对所需资源的相似性越低，资源竞争强度越小；如果基株与邻体之间的亲缘关系越近，则两者之间对所需资源的相似性越高，发生高强度资源竞争的可能性加大，导致光合作用速率下降。当光合作用速率下降到一定阈值，NSC 降低到一定程度时，碳收支失衡，树木则可能由于碳饥饿而死亡(McDowell 2011)。

从同资源种团水平树木存活分析上看，同种邻体指数对落叶树种树木的存活具有显著的负效应(图 3-1)，即当落叶树种基株周围的邻体中同种密度越高时，则基株的存活率越低，反之亦然，这说明该同资源种团群落存在显著的同种邻体负密度制约死亡效应。当基株周围的邻体中同种密度越高时，若基株与邻体具有相同的资源需求，则发生高强度资源竞争的可能性越大，导致光合作用速率下降，与群落水平树木死亡分析一样，树木则可能死于碳饥饿(McDowell 2011)。但是，无论是发生何种类型的负密度制约死亡效应，均对物种共存和生物多样性的维持产生正面效应。

3.5 本章小结

本章利用广义线性混合模型及似然比检验,探究了负密度制约效应对树木死亡的影响。结果表明,在小兴安岭凉水典型阔叶红松林,生物邻体因子对树木存活的效应存在显著的种间变异。但是,这种变异的检验结果受邻体半径的影响。树种叶习性可能是引起这种变异的原因之一。无论在固定 5 m、固定 10 m,还是在真实/近似的邻体半径的条件下,同种邻体指数对树木存活效应的种间变异幅度均大于异种邻体平均谱系相异性指数。因此,未来关于密度制约方面的研究,不应忽略邻体半径存在种间变异并对森林群落物种共存的研究结果产生影响的事实。稳定化机制中的负密度制约是小兴安岭凉水典型阔叶红松林物种共存和维持生物多样性的重要机理之一。

4 生境异质性对树木死亡的影响

4.1 引　　言

　　生境异质性是影响树木存活的一个重要的非生物因子，生境异质性与物种生境偏好引发的生境过滤作用是森林群落物种共存的重要机制之一(HilleRisLambers et al. 2012)。地形因子可能对树木存活的效应存在空间变异，可对土壤水分和太阳辐射进行重新分配，地形变异可能与物种生活史策略相互作用。例如，某些物种表现出对某类生境的偏好，然而另外一些物种则并不偏好此类生境(Druckenbrod et al. 2005, Messaoud and Houle 2006)。生境异质性和特定物种的生境偏好能够导致物种呈现空间聚集分布(Piao et al. 2013)。物种的生活史策略可能影响物种对外部胁迫因子(如光照和水分)的响应(Comita and Hubbell 2009, Gravel et al. 2010)。现已有大量研究表明，生境对物种分布及其生长和存活存在影响(Zhang et al. 2011, Metz 2012)。以往在温带与热带森林的研究结果表明，树种的空间分布与地形因子息息相关(Condit et al. 2010, Lan et al. 2011, 徐丽娜和金光泽 2012)。这可能是由于地形因子会影响地下水动态及光照条件空间变异的缘故(Harms et al. 2001)。

　　空间尺度可能影响地形因子对树木存活效应空间变异的检测。树木在不同空间尺度下，其空间分布格局存在变异(Chen and Bradshaw 1999, Wiegand et al. 2000, Schurr et al. 2004)，而此种变异与生境异质性或者斑块关系密切(Boyden et al. 2005, Riginos et al. 2005)。例如，在本研究区典型阔叶红松林的以往研究中发现，4 个主要冠层树种存在显著的空间聚集，但是这种聚集强度随着空间尺度的增大而下降，并且与地形因子存在密切联系(Liu et al. 2014)。生境异质性可能随着空间尺度的增大而增大，但是树种的聚集程度却可能随之下降。虽然以往有大量关于生境异质性与树木死亡方面的研究,例如，Wang 等(2012)在长白山阔叶红松林、Lu 等(2015)在八大公山亚热带常绿落叶阔叶混交林开展的研究，但是以往研究人员通常仅在单一空间尺度(树木个体尺度)下开展研究，在温带森林进行多重尺度下的此类研究甚少。

　　在本章中，使用 2005~2015 年共 10 年间隔的树木死亡动态监测数据集，探究生境异质性如何影响典型阔叶红松林树木死亡动态，并试图回答地形因子对树木死亡的效应是否存在显著的空间变异。主要解决以下 3 个小问题：①地形因子对树木死亡的效应是否存在显著的空间变异？②空间尺度能否对这种变异的检测造成影响？③生活史策略可否对上述变异做出合理解释？

相对应地提出以下 3 点假设：①地形因子对树木死亡的效应存在显著的空间变异；②空间尺度能够对这种变异的检测结果造成影响；③树种耐阴性、耐旱性之间的差异可能是上述变异的部分原因。

4.2 研究方法

4.2.1 数据收集

不同于树木生长和更新动态，树木死亡是一个相对缓慢的过程。为了尽可能研究较长时间段内的树木死亡动态，与第 3 章相同，本章选取 2005~2015 年共 10 年间的树木死亡动态监测数据集，参见 3.2.1 节。

4.2.2 样方尺度树木死亡驱动因子的构建

本章所采用的模型自变量包括 3 个样方尺度（5 m、10 m 和 20 m）下样方内树木初始平均胸径(average DBH, Avg DBH)、净亲缘关系指数(NRI)，以及海拔、凹凸度和山体阴影 3 个地形因子。由于第 3 章中已根据 AIC 模型拟合优度判断准则确定胸径、生物邻体与地形因子为最优变量组合，同时为了与第 3 章模型生境因子保持一致性，本章仅将地形因子视为非生物生境变量。

本章使用净亲缘关系指数来表示样方内物种亲缘关系(Webb et al. 2002)。该指数首先计算出样方中所有物种对的平均进化距离(mean phylogenetic distance, MPD)，为了排除群落间物种数不同所造成的误差，需要对 MPD 进行标准化，保持物种数量及物种个体数不变，将样方中的物种名在系统发育树上随机交换 999 次，进而获得该样方中物种在随机零模型下的 MPD 分布，最后利用随机分布结果将观测值标准化得到 NRI。

4.2.3 样方尺度树木死亡模型

本章借助于统计学中的广义地理加权回归模型(geographically weighted generalized linear model, GWGLM)及地理变异检验(geographical variability test, GVT)来试图解答地形因子对树木死亡的效应是否存在显著的空间变异(Fotheringham et al. 2003, Nakaya et al. 2009)。

关于地形因子对树木死亡的效应是否存在显著的空间变异的检验：采用广义线性模型(generalized linear model, GLM)和广义地理加权回归模型分别在 3 个样方尺度（5 m、10 m 和 20 m）下建立树木死亡模型。模型中因变量为每一个样方内 2005~2015 年共 10 年间树木死亡的株数。该因变量属于计数变量(count variable)，因此广义线性模型中采用 Poisson 回归分析方法，链接函数为对数函数。采用标准差标准化法(Z score)对所有自变量进行无量纲化处理,以便允许模型各个自变量参数估计值之间可以直接地比较大小。第3章中的树木存活个体尺度分析将单株树木视为研究对象,而在本章样方尺度上,则将样方内所有树木的集合作为研究对象。但一般 Poisson 回归模型并非完美。近年来,负二项回归模型(negative binomial, NB)、零膨胀 Poisson 回归模型(zero-inflated Poisson, ZIP)及零膨胀负二项回归模型(zero-inflated negative binomial, ZINB)逐渐应用于因变量为计数变量的统计模型建立当中。负二项回归模型考虑了因变量可能存在过度离势(overdispersion)的问题。实际上,Poisson 分布是负二项分布的一个特例。零膨胀 Poisson 回归模型充分考虑了因变量可能出现零膨胀的问题,即因变量中零值过多的现象。零膨胀负二项回归模型综合了上述两类模型。负二项回归模型、零膨胀 Poisson 回归模型及零膨胀负二项回归模型均为对一般 Poisson 回归模型的补充和完善。因此,本章在构建全局模型(global model)时,除了采用一般 Poisson 回归模型,也同样应用了负二项回归模型、零膨胀 Poisson 回归模型及零膨胀负二项回归模型。Vuong 检验(Vuong's test)是一种对非嵌套统计模型进行模型比较的统计量,可以对上述4种全局模型的拟合优度进行评价比较(Vuong 1989)。

分别计算3个空间尺度下的全局空间自相关指数(Moran's index),发现在任何一个样方尺度下,因变量均呈现出了显著的空间自相关性并表现出空间聚集,参见4.3.1节。因变量具有显著的空间自相关性是应用地理加权回归模型技术的前提条件,为模型的构建奠定了基础。因此,将2005~2015年共10年间样方内树木死亡株数设置为广义地理加权回归模型的因变量具有充分可靠的理论依据。本研究将广义线性模型视为全局模型(global model),而将广义地理加权回归模型视为局域模型(local model)。在全局模型中,模型所有自变量参数估计值在样地的任何空间位置上均保持不变。但在局域模型中,模型的某些参数估计值在空间位置上存在差异,换言之,某些参数估计值不是固定不变的,而是视空间位置而变。本研究采用校正后的赤池信息准则(AICc)来比较全局模型与对应的局域模型的优劣。

在进行广义地理加权回归模型参数估计时,采用高斯核函数计算空间权重矩阵(Fotheringham et al. 2003, Guo et al. 2008)。应用间隔搜索法来确定最优带宽以确保稳健。由于 AICc 通常被认为是最好的拟合泊松回归最优指标(Burnham and Anderson 2002),因此最优带宽的判别标准采用 AICc(Nakaya et al. 2005)。其他有关地理加权回归模型的详情可参考文献(Fotheringham et al. 2003)。

本研究利用地理变异检验,判定地形因子对树木死亡的效应是否存在显著的空间变异,即判定一般地理加权回归模型(original GWR)与半参数化地理加权回归模型(semiparametric GWR)之间的差异是否显著。地理变异检验用来检验某个自变量参数估计值在地理空间位置上是否存在显著的空间变异。通过对比一般地理加权回归模型与半参数化地理加权回归模型来完成。半参数化地理加权回归模型中只有第 k 个参数为全局参数，而其余参数均为局域参数（即与一般地理加权回归模型的参数一致），根据 AICc 等模型拟合优度比较准则，如果一般地理加权回归模型优于半参数化地理加权回归模型，则可以确定第 k 个参数在地理空间上存在显著的空间变异。

空间尺度可能影响地形因子对树木死亡效应空间变异的侦测。本研究在 3 个样方尺度（5 m、10 m 和 20 m）下分别通过比较一般地理加权回归模型和半参数化地理加权回归模型，以及借助地理变异检验来判定空间尺度能否影响地形因子对树木存活效应空间变异的侦测。

广义线性模型构建如下：

$$\ln(\mu) = \sum_k (\beta_k \cdot x_k) + \varepsilon \tag{4-1}$$

式中，μ 代表预测的样方内树木死亡个体数；β_k 代表第 k 个自变量的参数估计值；x_k 代表第 k 个自变量。

广义地理加权回归模型构建如下：

$$\ln(\mu(u_i, v_i)) = \sum_k [\beta_k(u_i, v_i) \cdot x_{ki}] + \varepsilon_i \tag{4-2}$$

式中，$\beta_k(u_i, v_i)$ 代表第 i (u_i, v_i) 个空间位置上（即某一样方位置）第 k 个自变量的参数估计值。

一般广义地理加权回归模型和半参数化广义地理加权回归模型构建如下（以海拔的参数估计值为例）。

一般广义地理加权回归模型：

$$\begin{aligned}\ln(\mu(u_i, v_i)) = &\beta_0(u_i, v_i) + \beta_1(u_i, v_i) \cdot \text{AvgDBH}_i + \beta_2(u_i, v_i) \cdot \text{NRI}_i + \beta_3(u_i, v_i) \\ &\cdot \text{Elevation}_i + \beta_4(u_i, v_i) \cdot \text{Convexity}_i + \beta_5(u_i, v_i) \cdot \text{Hillshade}_i + \varepsilon_i\end{aligned} \tag{4-3}$$

半参数化广义地理加权回归模型：

$$\begin{aligned}\ln(\mu(u_i, v_i)) = &\beta_0(u_i, v_i) + \beta_1(u_i, v_i) \cdot \text{AvgDBH}_i + \beta_2(u_i, v_i) \cdot \text{NRI}_i + \beta_3 \\ &\cdot \text{Elevation}_i + \beta_4(u_i, v_i) \cdot \text{Convexity}_i + \beta_5(u_i, v_i) \cdot \text{Hillshade}_i + \varepsilon_i\end{aligned} \tag{4-4}$$

与第 3 章所采用的广义线性混合模型类似，本章采用的广义地理加权回归模型也属于广义线性模型的一种扩展。两类广义线性模型存在着一些异同点。其中，相同点为两种模型都允许自变量参数估计值发生变化，即斜率是可以变化的。此外，在探讨两类模型参数是否存在显著变异时，广义线性混合模型应用了似然比检验法，而广义地理加权回归模型应用了地理变异检验法，两种方法虽然各异，但是原理非常类似，均是通过对各自模型分别建立的两个亚模型的方式进行比较判断。对于广义线性混合模型而言，两个亚模型分别为随机截距模型与随机系数模型。而对于广义地理加权回归模型而言，两个亚模型分别为一般广义地理加权回归模型与半参数化广义地理加权回归模型。不同点在于广义线性混合模型中的亚模型之一——随机系数模型允许随机效应属于任何数据类型，包括空间与非空间数据类型。例如，树种和样方号可作为随机效应，模型通用性较强。虽然广义地理加权回归模型只能允许变量的斜率在不同地理空间位置上发生变化，但是它能够更有效地处理模型误差中的空间自相关和空间异质性问题（Zhang and Gove 2005, Zhang et al. 2009, Zhen et al. 2013），使得该模型专用性较强。因此，这两类广义线性模型并无优劣之分，各有擅长之处，均为对广义线性模型的补充。

同资源种团/功能群（guild/functional group）分析：与第 3 章类似，为了解释地形因子对树木死亡的效应存在显著的空间变异的原因，本章根据树种生活史策略及其资源利用方式将树种分为不同的同资源种团。例如，根据耐阴性将树木分为耐阴树种、中性树种和喜光树种 3 个同资源种团；根据耐旱性将树木分为湿生树种、中生树种和旱生树种 3 个同资源种团。分别为每一个同资源种团在个体尺度下建立树木存活模型，得到各个模型参数估计值的大小、正负和显著性，通过分析比较不同同资源种团模型之间的差异，探讨生活史策略对上述空间变异的影响。此分析模型细节参见 3.2.3 节。

广义线性模型、广义地理加权回归模型及地理变异检验过程均采用 GWR4 软件实现（Fotheringham et al. 2003）。负二项回归模型采用 R 软件 MASS（Venables and Ripley 2002）程序包拟合。零膨胀 Poisson 回归模型及零膨胀负二项回归模型、Vuong 检验采用 R 软件 pscl（Zeileis et al. 2008）程序包拟合。负二项回归模型、零膨胀 Poisson 回归模型及零膨胀负二项回归模型参数估计采用极大似然估计法（maximum likelihood estimation, MLE）。广义地理加权回归模型的参数估计采用加权最小二乘法（weighted least squared, WLS）（Fotheringham et al. 2003）。GWR4 是一个界面友好的地理加权回归建模专用软件。目前，能实现地理加权回归模型的软件虽然较多，如 R 软件的程序包 spgwr（Bivand and Yu 2015）、gwrr（Wheeler 2013）、GWmodel（Gollini et al. 2015）及地理信息系统 ArcGIS 10.1（ESRI, Redlands, CA, USA）等软件，但是前述各个程序包和软件的当前版本仅能实现基本的地理加权回归或者广义地理加权回归模型，尚未增加对半参数化地理加权回归模型及地理变异检验的支持。因此，本研究采用 GWR4 软件拟合各类地理加权回归模型及其相关检验，该软件最新版本支持了半参数化地理加权回归模型及地理变异检验功能。

4.3 结　　果

4.3.1 样方尺度树木死亡的地形驱动因子

分别在 5 m、10 m 和 20 m 样方空间尺度下建立模型。结果表明，在上述任何空间尺度下，任何一个局域模型均优于与它相对应的全局模型(表 4-1)。在 10 m 样方尺度下，所有的地形因子效应均存在显著的空间变异(表 4-2)。样方内死亡木数量的全局空间自相关指数在 5 m、10 m 和 20 m 尺度下分别为 0.2100($P<0.001$)、0.3218($P<0.001$) 和 0.2694($P<0.001$)。随着样方尺度树木死亡模型中因变量空间自相关性的增大，参数具有显著空间变异的地形因子的数量也随之增加(表 4-2)。

表 4-1　不同空间尺度下广义线性模型及其对应的广义地理加权回归模型之间的模型比较
Table 4-1　Modal comparison between global (GLM) and local (GWGLM) models at different spatial scales

空间尺度	模型类型	校正赤池信息准则	被解释的偏常比例/%
5 m 样方	全局模型	3119	14.51
	局域模型	**3003**	**25.88**
10 m 样方	全局模型	1427	29.18
	局域模型	**1366**	**35.38**
20 m 样方	全局模型	379	54.79
	局域模型	**324**	**70.06**

注：最优模型的校正赤池信息准则和被解释的偏常比例为粗体显示。

表 4-2　不同空间尺度下广义地理加权回归模型的地理变异检验
Table 4-2　Geographical variability tests of local coefficients of GWGLM at different spatial scales

空间尺度	变量	AICc 差异
5 m 样方	平均胸径	7.4753
	净亲缘关系指数	2.1287
	海拔	9.6838
	凹凸度	10.5408
	山体阴影	9.3248
10 m 样方	平均胸径	−10.0464
	净亲缘关系指数	0.7054
	海拔	−7.1549
	凹凸度	−3.1340
	山体阴影	−1.0224

续表

空间尺度	变量	AICc 差异
20 m 样方	平均胸径	−31.0416
	净亲缘关系指数	9.4754
	海拔	−2.7442
	凹凸度	5.3192
	山体阴影	5.9047

注：校正后赤池信息准则为负值时，说明其对应的自变量参数估计值存在显著的空间变异。

此外，个体尺度树木存活分析结果表明，地形因子对不同同资源种团树木存活的效应存在明显差异。以海拔为例，海拔对喜光树种树木的存活为显著正效应，但对耐阴树种树木的存活则为负效应且不显著(图 4-1)。海拔对旱生树种树木存活

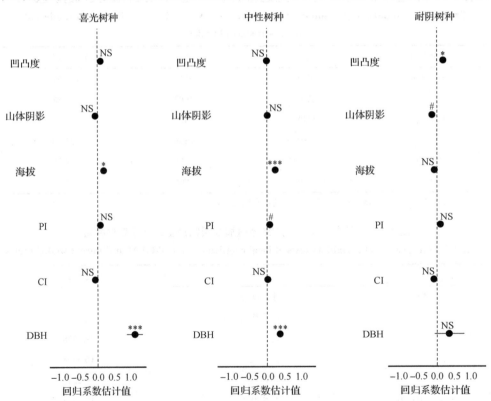

图 4-1　树木大小、生物邻体及地形因子对喜光树种、中性树种、耐阴树种树木存活的影响

Fig. 4-1　Standardized parameter estimates (± SE) of the effects of tree size, biotic neighbourhood and topographic variables on tree survival of shade-tolerant, mid-tolerant and shade-intolerant functional groups

存在显著正效应,但对湿生树种树木存活存在显著负效应(图 4-2)。海拔比凹凸度和山体阴影对树木存活有更显著的效应。例如,除耐阴树种外,海拔对其他几种同资源种团(中性、耐阴、湿生、中生和旱生)的树木存活均为显著的效应(图 4-1、图 4-2)。

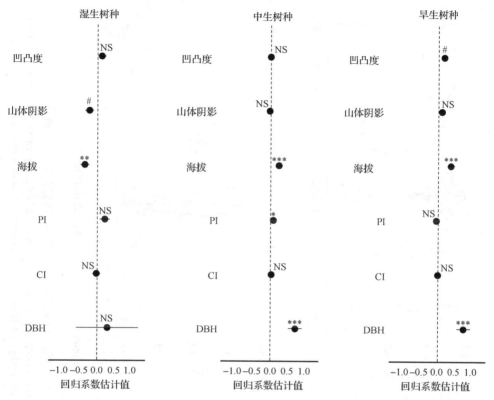

图 4-2 树木大小、生物邻体及地形因对湿生树种、中生树种、旱生树种树木存活的影响
Fig. 4-2 Standardized parameter estimates (± SE) of the effects of tree size, biotic neighbourhood and topographic variables on tree survival of hygrophilous, mesophilous and xerophilous functional groups

4.3.2 不同空间尺度下地形效应空间变异检验

综合参数估计值分布与方差可得,无论在任何一个空间尺度下,海拔效应的空间变异幅度均大于凹凸度和山体阴影效应的空间变异幅度(图 4-3、表 4-3)。此外,在任一样方尺度、任一类型的全局模型中,海拔均为对树木死亡影响最大并且显著的唯一地形因子(图 4-4)。

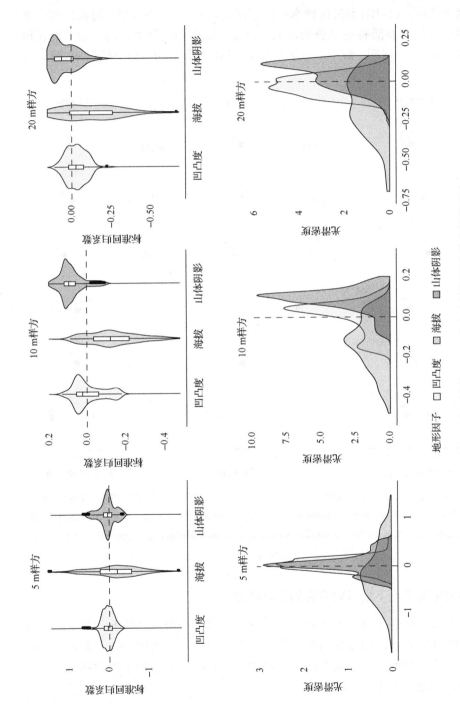

图4-3 不同空间尺度下海拔、凹凸度和山体阴影参数估计值分布

Fig. 4-3 Violin plot and smooth density plot showing standardized parameter estimates of effects of elevation, convexity and hillshade on tree mortality at different quadrat scales

4 生境异质性对树木死亡的影响

表 4-3 不同空间尺度下海拔、凹凸度和山体阴影回归系数估计值的方差

Table 4-3 Variance of coefficients of elevation, convexity and hillshade at different spatial scales

空间尺度	海拔	凹凸度	山体阴影
5 m 样方	0.3483	0.0214	0.0224
10 m 样方	0.0188	0.0072	0.0020
20 m 样方	0.0408	0.0052	0.0086

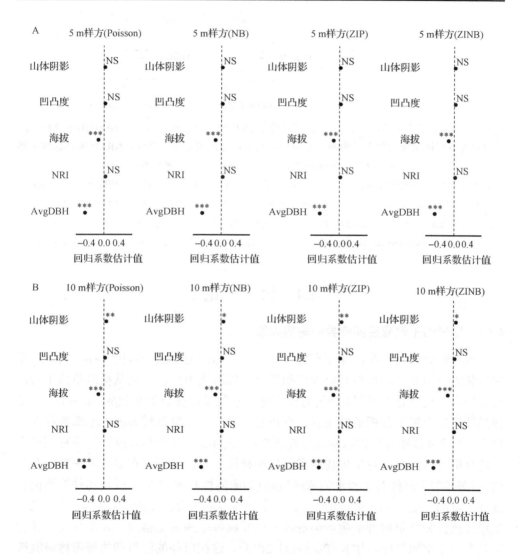

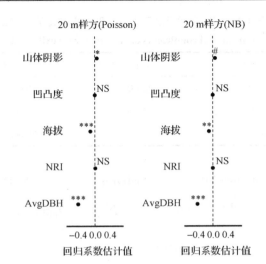

图4-4 全局水平下平均胸径、净亲缘关系指数及地形因子分别在5 m(A)、10 m(B)和20 m(C)样方尺度下对树木死亡数量的影响，AvgDBH代表样方内树木平均胸径；NRI代表净亲缘关系指数（根据Vuong检验，负二项回归为最优模型形式）

Fig. 4-4 Standardized parameter estimates (± SE) of the effects of AvgDBH, NRI and topographic factors on mortality at sub-quadrat (A: 5 m), standard-quadrat (B: 10 m) and index-quadrat (C: 20 m) scales at the global level. AvgDBH refers to average DBH of individuals within a quadrat; NRI refers to the net relatedness index. Negative binomial regression is the best structure via Vuong's test

4.4 讨 论

4.4.1 地形因子效应空间变异的影响因素

本章研究结果表明，在凉水典型阔叶红松林，地形因子对树木存活的效应存在显著的空间变异，这与研究假设相符合。以海拔为例，虽然从样地整体上看，按全局水平模型分析可得，海拔对样方死亡木数量存在显著负效应（图4-4），但是按局域尺度上模型分析的结果看，海拔对大多数空间位置样方死亡木株数存在负效应，但对少数样方死亡木株数存在正效应（图4-3）。从个体尺度同资源种团树木存活分析可得，海拔对旱生树种和湿生树种树木存活分别存在显著的正效应及负效应（图4-2）。树种对光和水的适应策略（如耐阴性和耐旱性），以及物种生境偏好及其引发的生境过滤作用，可能是这种空间变异的主要驱动因素（图4-1、图4-2）。生境异质性和特定树种的生境偏好能够引发树种空间聚集格局，使得特定树种被限制在一定空间范围内生长（Piao et al. 2013），这在以往的温带和热带森林研究树种空间分布特征中多有证实，如小兴安岭典型阔叶红松林、西双版纳热带雨林、巴拿马BCI热带雨林、泰国HKK热带雨林和马来西亚Pasoh热带雨林动态监测样地(Condit 2010, Lan et al. 2011, Lan et al. 2012)。徐丽娜和金光泽(2012)发现小兴安岭典型阔叶红松林主要树种的空间分布呈现聚集态势，并且树种的空间分布格局与地形因子密切相关；Lan 等(2012)发现西双版纳热带雨林 20 个主要树种当中，有 14 个树种的空间分布与某一个或某两个地形因子显著相关。若干温带与热带的森林动态监测样地，如长白山温带森林、厄瓜多尔 Yasuní 低地热带雨林动态监测样地，树种的空间分布格局、树木生长和存活与其所处生境密切相关(Zhang et al. 2011, Metz 2012)。

海拔对树木死亡效应的空间变异幅度大于凹凸度和山体阴影(图 4-3、表 4-3)。此外，在个体尺度下群落水平上树木存活模型及样方尺度下树木死亡全局模型当中，海拔均为对树木死亡影响最大并且显著的地形因子(图 3-2、图 4-4)。在同资源种团分析中，海拔比凹凸度和山体阴影对树木存活有更显著的效应(图 4-1、图 4-2)。湖南省八大公山亚热带常绿落叶阔叶混交林 25 hm^2 动态监测样地(高差约 100 m)研究发现，在特征组水平上，海拔对极常见种、常见种及成年树组的死亡存在显著影响，但是其他地形因子对树木死亡无显著影响(Wu et al. 2017)。对长白山阔叶红松林 25 hm^2 动态样地(高差＜18 m)研究发现，在同资源种团水平上，海拔仅对极稀有种组树木死亡存在显著影响(Wang et al. 2012)。可见，在某些条件下，海拔对树木死亡存在一定的影响。但是，这并不能推出海拔是树木死亡的直接影响因素，因为海拔属于地形因子，是一种间接生态因子，只能通过与它相关性较高的其他因素间接对树木死亡造成影响。在本研究区样地内，海拔虽然变异幅度不大(高差约 80 m)，但是某些土壤水分和养分等理化属性变异幅度较大，而且海拔与土壤属性之间相关性较高，本研究区海拔与土壤第二主成分轴之间存在较强的相关性($r = 0.48$, $P＜0.001$)。由此可见，海拔可能与某些土壤属性发生协变，并间接地影响树木死亡，真正直接影响树木死亡的因素可能是某些土壤属性。再者，样地整体地势特征明显，大体呈现东高西低、北高南低，海拔较高地区往往处于山坡和山脊处，而海拔较低地区则位于山谷处(图 2-1)，样地海拔可能与坡位关系密切，而某些树种生境偏好同样与坡位关系密切。例如，红松侧根发达，偏好生长在土壤排水良好的陡坡地区；而臭冷杉则耐水湿，偏好生长在地势平坦区域。海拔可能通过与它相关性较高的其他因素(如某些变异幅度较大的土壤属性)发生协变，间接地影响树木死亡。西双版纳热带雨林动态监测样地的研究显示，相较于凹凸度、坡度和坡向，海拔是影响树种空间分布最重要的地形因子，海拔与土壤含水率呈显著相关(Lan et al. 2011)。

本章得出的地形因子对树木存活的效应存在显著的空间变异这一结论仍然存在一定的局限性。首先，本章中所应用的时空尺度对于检验这些关系可能仍然相对较小。海拔效应一般在更大的空间尺度上更加明显。本章在 10 m 样方尺度上捕获到最大的地形效应变异，但这也有可能是受到所采用的粒度(grain)和幅度

(extent)影响所致。尺度对空间格局分析的影响可能随着粒度和(或)幅度的变化而发生变化(Wu 2004)。在不同空间尺度上开展研究或者改变粒度和幅度可能会对探索空间格局有更好的帮助(Wiens 1989)。其次，由于目前的 GWR 模型软件或相应的 R 软件程序包均尚未能实现 GWR 与负二项回归模型、零膨胀 Poisson 回归模型及零膨胀负二项回归模型的集成，因此本章在局域水平模型，即 GWR 模型中只采用了一般 Poisson 回归模型。随着 R 软件程序包的迅速发展和日臻完善，在以后的研究中，可期待在 GWR 模型中集成负二项回归模型、零膨胀 Poisson 回归模型及零膨胀负二项回归模型形成复杂模型软件包。但是，无论采用上述何种类型的全局水平模型并且在何种样方尺度下，海拔对树木死亡数量的影响均为显著负效应，且在所有地形因子中对树木死亡影响最大(图 4-4)。

4.4.2 空间尺度对检验地形因子效应空间变异的影响

空间尺度可能会影响地形因子对树木死亡效应空间变异的检测，这与研究假设相符。本章研究发现，只有在 10 m 样方尺度下才能检测到所有地形因子的参数估计值均存在显著的空间变异(表 4-2)。随着样方尺度分析模型中因变量空间自相关性的增大，参数具有显著空间变异的地形因子的数量也随之增加(表 4-2)。Liu 等(2014)研究发现，小兴安岭凉水典型阔叶红松林 4 个主要冠层树种空间分布均存在显著的空间聚集格局，然而这种空间聚集程度会随着空间尺度的增大而下降，并且地形被认为是影响该区域这些树种空间分布的主要生态因子(Liu et al. 2014)。生境异质性可能随着空间尺度由微尺度(如 5 m×5 m)向广尺度(如 20 m×20 m)变化而增大，然而物种的空间聚集程度可能会随着空间尺度由微尺度向广尺度变化而下降。实际上，生境变量的计算是受空间尺度(即分辨率或样方大小)的影响。因此，地形对树木存活效应的空间变异可能受到空间尺度制约(scale dependent)。

4.4.3 生境异质性导致的树木死亡

在个体尺度下群落水平树木存活分析中，海拔对树木存活存在显著正效应(图 3-2)，即树木所属样方的海拔越高，则该树木的存活率越高，反之亦然。结合样地实地踏查和以往研究结果分析，这可能是由于样地内分布着大量的大径级、长寿命的红松导致分析数据集中这部分的样本量较多，而大多数红松在生活史早期阶段已经经历了生境过滤并且成功地存活下来，在其生长的生境条件具有良好的适应性，并生长在环境适宜地区，如陡坡地区，而样地内的几处坡度较陡的地区往往海拔相对较高(图 2-1)。红松侧根发达，喜好生长在排水良好的陡坡地区。因此，从群落水平上看，2005～2015 年共 10 年间，海拔对树木的存活存在着显著的正效应。

从同资源种团水平树木存活分析上看，地形因子对不同耐阴性、耐旱性树种的存活存在着明显不同的影响(如参数值大小、正负和显著性)(图 4-1、图 4-2)。这是由于不同的同资源种团或不同树种资源利用方式差异所致。例如，海拔对湿生树种的存活存在着显著的负效应。样地的地势总体特征为东高西低，虽然海拔高差只有大约 80 m，但是，土壤水分和养分等理化属性变异幅度较大，而且海拔与土壤属性之间相关性较高，土壤属性随海拔发生协变。若湿生树种生长在土壤水分含量相对较低的高海拔地区，则对湿生树种树木的存活产生不利影响。生长在这一地区的湿生树种可能由于缺少充足的水分而引发水力失衡，这可能是导致样地内海拔越低的地区湿生树种死亡率越高的原因。

4.5 本章小结

本章应用广义地理加权回归模型及地理变异检验，探究了生境异质性对树木死亡的影响。结果表明，在小兴安岭凉水典型阔叶红松林，地形因子对树木死亡的效应存在显著的空间变异。但是，这种变异的检验结果受空间尺度的影响。树种对光和水分环境的适应策略(即耐阴性与耐旱性)可能是引起上述变异的部分原因。在 5 m、10 m 及 20 m 样方空间尺度条件下，海拔对树木死亡效应的空间变异幅度均大于凹凸度和山体阴影。未来研究应该考虑空间尺度变化对森林群落物种共存研究结果的影响。

5 树木死亡-生长种间权衡与物种共存

5.1 引　言

不仅负密度制约与生境异质性可通过对树木死亡的影响促进物种共存并维持森林生物多样性，而且树木死亡与生长通过树木死亡-生长种间权衡也可对促进物种共存具有一定贡献(Kitajima and Poorter 2008)。森林动态与森林群落构建通常受到森林死亡与生长动态种间变异的影响(Pacala et al. 1996, Rees et al. 2001, Russo et al. 2010)。此外，树木死亡概率与树木生长效率(生长活力)之间的关系同样非常密切(Lin et al. 2017)。因此，树木死亡、生长及两者间的权衡关系在森林动态中发挥着不可忽视的作用，它们是多个生态系统过程的基础，并促进森林群落物种共存(Lutz and Halpern 2006, Kitajima and Poorter 2008, Iida et al. 2014b)。研究树木死亡、生长及两者之间的关系，有助于生态学家更好地理解森林群落构建与动态。

与树木死亡类似，树木生长也受生物和非生物因子综合影响，并且生物和非生物因子对树木死亡与生长的影响随着生活史阶段而变化(Visser et al. 2016)。近年来温带与热带森林的研究结果显示，同种邻体密度被认为是幼苗存活、存续(persistence)与更新，以及树木生长、存活最重要的生物驱动因子(Comita et al. 2010, Johnson et al. 2012, Wang et al. 2012, Zhang et al. 2016b, Du et al. 2017, Lin et al. 2017)。异种邻体功能性状相异性或谱系相异性对基株的影响，同样也被认为是幼苗存活、树木存活与生长的另一个重要的生物邻体驱动因子(Webb et al. 2006, Paine et al. 2012, Lebrija-Trejos et al. 2014)。以往温带与热带森林的研究中发现，生境变量(如地形、光照、水分、土壤属性等)通过生境过滤作用，与生物邻体因子共同影响着幼苗生长、存续及树木生长与存活(Wang et al. 2012, Chi et al. 2015, Zhang et al. 2016b, Lin et al. 2017)。以往大多数研究均只关注局部环境因子对某一项森林动态(如生长或存活)的影响，并基于此进一步探讨了物种共存与群落构建机制(Wang et al. 2012, Piao et al. 2013, Johnson et al. 2014, Chi et al. 2015, Zhu et al. 2015b, Chen et al. 2016)。然而，局域因子对树木死亡与生长的影响是否存在差异尚不清晰。

树木死亡与生长之间的种间权衡关系最能体现树种生活史策略差异(Grubb 1977, Hubbell and Foster 1992, Pacala et al. 1996, Wright et al. 2003, Stephenson et al. 2011)。通常树木生长与死亡动态种间差异更容易发生在小树阶段，这在热带森林研究中已得到广泛证实(Kitajima 1994, Wright et al. 2003, Poorter et al. 2008,

Wright et al. 2010, Iida et al. 2014a, Iida et al. 2014b)。再者，树木死亡与树木年直径生长量存在关联。通常，树木年直径生长量在小树阶段表现出上升趋势，而在大树阶段表现出下降趋势(随着树木的衰老，树木生长率放缓)。随着树龄的增长，当树木受到机械损伤、昆虫或病原菌攻击等外源伤害时，树木的生长受到更大的削弱，进而使树木死亡率增高。若树木年生长量不能维持某一最小值，那么树木的死亡风险就会增大(Botkin et al. 1972)，树木生长效率的下降将可能导致树木存活率下降、死亡率上升(Leemans 1991)。虽然近年来关于树木生长与死亡关系的研究较多(Cailleret et al. 2017)，然而在温带森林同时开展树木生长与死亡种内、种间关系的研究却鲜有报道。

在本章中，使用2010~2015年共5年间的树木生长与死亡动态监测数据，探究树木死亡-生长种间权衡与物种共存机制(稳定化机制与均等化机制)之间存在何种关系，并试图回答以下两个问题：①在群落水平上，局域因子对树木死亡与生长的影响是否存在差异？是否同时存在负密度制约死亡与生长效应？②树木死亡与生长之间存在何种种内与种间关系？是否因此而促进物种共存？

相对应地提出两点假设：①在群落水平上，局域因子对树木死亡与生长的影响存在差异，同时存在负密度制约死亡与生长效应；②树木死亡与生长在种内存在负相关关系，而在种间存在正相关关系，即权衡关系，从而促进物种共存。

5.2 研究方法

5.2.1 数据收集

与树木死亡动态的缓慢性不同，树木的生长行为无时无刻不在进行当中。另外，为了尽可能研究具有较高完整度生活史阶段的树木死亡和生长动态，本章选取2010~2015年共5年间的树木死亡和生长动态监测数据集。因为这一时间段的数据包括了幼树1~2 cm相对较早的生活史阶段，即在2010年，对样地内所有DBH≥1 cm的活的木本植物进行编号挂牌、测量胸径、定位坐标并鉴别树种。在2015年，对2010年所调查的活立木进行复查，记录其胸径和存活状态(存活或死亡)。选取40株具有完整3个功能性状(比叶面积、最大树高及木质密度)的树种作为原始数据集来建立生长数据集和死亡数据集。在建立生长数据集时，我们剔除以下情况：①剔除主干损坏而用萌生枝代替测量的；②根据以往研究并结合温带森林特征，剔除年生长量大于2 cm或者年萎缩量大于初始胸径25%以上者；这里并未采用热带森林的标准，因为通常温带森林树木年直径生长量大大低于热带森林；③为了精确计算年直径生长量，即绝对生长率，剔除了2010~2015年的死亡木数据。虽然不把这些剔除的树木作为基株，但是仍然将剔除的树木作为邻体来处理，以便为其他基株计算生物邻体因子所用。地形数据收集参见2.2.2节。

5.2.2 局域驱动因子的构建

本章树木存活与生长模型的自变量包括：树木初始胸径大小、生物邻体与地形因子。由于第 3 章中已根据 AIC 模型拟合优度判断准则确定胸径、生物邻体与地形因子为最优变量组合，同时为了与第 3 章模型生境因子保持一致性，本章仅将地形因子视为非生物生境变量。树木大小是一个非常重要的强烈影响树木死亡与生长的内生变量(King et al. 2006, Wang et al. 2012, Zhang et al. 2016b)。

树木生长与死亡的外部驱动因子包括生物和非生物驱动因子。生物因子包括同种邻体指数和异种邻体平均性状相异性指数(average trait dissimilarity index of heterospecific neighbor, TI)(Paine et al. 2012, Chen et al. 2017)。这两个指数为基于邻体组成、初始胸高断面积、坡位、功能性状及高斯核函数计算得出(Pu et al. 2017)。异种邻体平均性状相异性指数越大，则表示基株与邻近个体之间的性状差异越大。由于第 3 章中已根据 AIC 模型拟合优度判断准则确定胸径、生物邻体与地形因子为最优变量组合，为了与第 3 章模型非生物因子保持一致，本章仅将地形因子视为非生物变量。

植物功能性状是指对植物体定植、生长和死亡存在潜在影响的一系列植物属性，且这些属性能够单独或联合指示生态系统对环境变化的响应，并且能够对生态系统过程产生强烈影响。近年来，也有学者在树木死亡与生长模型中采用功能性状构建生物邻体因子(Lugo and Scatena 1996)。Chen 等(2016)认为邻体性状差异与树木生长关系密切。为了探究功能性状对树木死亡和生长的影响，同时也为了与前人研究对比，本章在构建生物邻体因子时采用了异种邻体平均性状相异性指数，代替了第 3 章中的异种邻体平均谱系相异性指数。实际上，异种邻体平均性状相异性指数与异种邻体平均谱系相异性指数比较类似，尤其是在植物功能性状具有系统发育保守性(phylogenetic niche conservatism)的前提假设下(Webb et al. 2002)。

同种邻体指数公式定义见公式(3-1)，高斯核空间距离权重函数定义见公式(3-3)，异种邻体平均性状相异性指数公式定义如下：

$$\mathrm{TI} = \sum_i (\mathrm{TD}_i \cdot \mathrm{HBA}_i \cdot W_i) / N \tag{5-1}$$

式中，TI 代表异种邻体功能性状相异性指数；HBA_i 为异种邻体胸高断面积(m^2)；i 代表树牌号；W_i 代表高斯核空间距离权重函数；TD_i 代表基株与邻体之间的功能性状空间内的欧氏距离；N 代表异种邻体株数。

5.2.3 树木死亡与生长模型

树木存活模型的构建过程与第 3 章相同，参见 3.2.3 节。树木生长模型采用高斯回归形式的广义线性混合模型(Gaussian GLMM)。高斯回归形式的广义线性混合模型实际上即为狭义的线性混合模型。因变量为树木的绝对直径生长率(absolute growth rate, AGR)(Chen et al. 2016)，且服从高斯分布。由于在分析中发现35 cm是一个明显的转折点(详见5.3.2节)，因此以胸径35 cm为阈值划分小树和大树，并且对各组树木分别建立个体尺度下树木存活模型和生长模型。

5.2.4 树木死亡与生长关系检验

树木死亡与生长种间关系检验：本章采用Pearson相关性分析检验树木生长率与死亡率之间是否存在显著的种间权衡关系。为了表征树木生长并且与以往研究保持一致(Wright et al. 2010)，本章构建3个相对生长率(relative growth rate, RGR)指标，即适宜条件下的相对生长率(RGR_{95}、RGR_{90})和平均相对生长率($RGR_{average}$)。死亡也分别构建3个相应的死亡率(mortality rate, MR)，即不适宜条件下的死亡率(MR_{25}、MR_{50})和总体死亡率($MR_{overall}$)。相对生长率是指本年实测胸径值除以上一次调查时的胸径值经自然对数转换后再除以间隔年份；死亡率是指每个树种调查间隔期内死亡个体数除以上一次调查时的总个体数(Wright et al. 2010)。适宜条件下的相对生长率(RGR_{95}、RGR_{90})等于每个树种第95或第90分位数上的相对生长率(每个树种的树木按其相对生长率降序排列)；不适宜条件下的死亡率(MR_{25}、MR_{50})是指之前调查间隔中相对生长率最慢的25%或50%个体的死亡率(Wright et al. 2010)。以上各类生长率和死亡率指标，仅对树种多度至少达到100的树种进行计算。

树木死亡与生长种内关系检验：本章还按照不同径级将树木划分为9个组(阈值：<5 cm、5~10 cm、10~20 cm、20~30 cm、30~40 cm、40~50 cm、50~60 cm、60~70 cm和>70 cm)，分别计算每个分组的树木死亡比例和平均绝对生长率，用以评价树木死亡与生长之间的种内关系。

广义线性混合模型均采用R 3.1.3软件中的"lme4"(Bates et al. 2015)程序包拟合。

5.3 结　　果

5.3.1 局域因子对树木死亡与生长的影响

在群落水平上，局域因子对树木死亡与生长存在差异，但差异性并不明显(图5-1、图5-2)。生物邻体因子CI对树木存活和生长均存在显著负效应，生物邻体因子TI对树木生长存在显著正效应，但对树木存活的效应则不显著；海拔和凹凸度对树木的存活和生长均存在显著或边缘显著的正效应；山体阴影则对树木的存活和生长均无显著效应(图5-1、图5-2)。

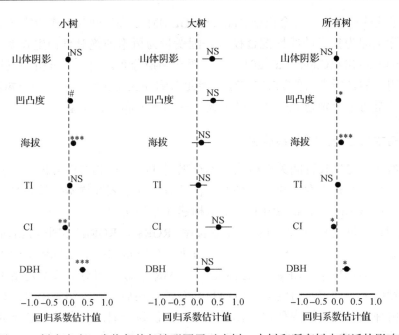

图 5-1 树木大小、生物邻体与地形因子对小树、大树和所有树木存活的影响
Fig. 5-1 Standardized parameter estimates (± SE) of tree size, biotic neighborhood and topographic variables on tree survival for small and large trees (i.e., DBH size cut-off of 35 cm) and all trees

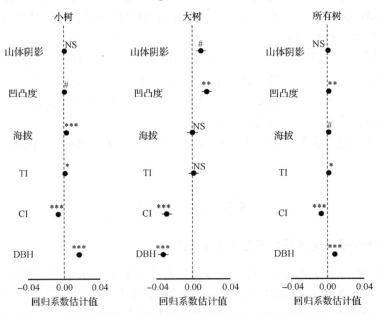

图 5-2 树木大小、生物邻体与地形因子对小树、大树和所有树木生长的影响
Fig. 5-2 Standardized parameter estimates (± SE) of tree size, biotic neighborhood and topographic variables on tree growth for small and large trees (i.e., DBH size cut-off of 35 cm) and all trees

5.3.2 树木死亡与生长的种内及种间关系

树木初始胸径大小对小树和所有树木的存活存在显著正效应，但是对大树的存活，其效应并不显著(图 5-1)。树木初始大小对小树和所有树木的生长具有显著正效应，而对大树的生长具有显著负效应(图 5-2)。此外，随着径级增大，每个径级的树木平均绝对生长率出现先升高、随后平缓、最后明显下降的趋势(图 5-3)。随着径级增大，每个径级内树木死亡比例首先出现急剧下降趋势，随后趋于平缓(图 5-3)。通过分析发现，35 cm 可能是一个个体发育的转折点(ontogenetic turning point)，在这个转折点上死亡和生长两条曲线的斜率同时接近 0(图 5-3)。

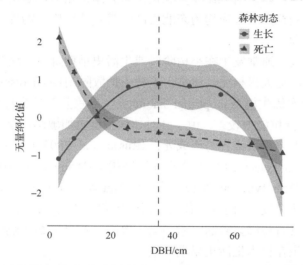

图 5-3 不同径级下的树木死亡比例和平均绝对生长率(经过无量纲化处理)

Fig. 5-3 Dimensionless tree mortality proportion and mean absolute diameter growth rate within each DBH size classes

在小树和所有树中，RGR_{90} 与 MR_{50} 之间为显著正相关($r=0.49$, $P=0.03$; $r=0.50$, $P=0.02$)，说明两者存在显著的树木生长与死亡种间权衡关系，但大树则不存在显著关系；RGR_{95} 与 MR_{25} 之间、$RGR_{average}$ 与 $MR_{overall}$ 之间均呈不显著正相关(表 5-1)。

表 5-1 树木生长和死亡种间权衡关系检验

Table 5-1 Trade-off relationship test between tree growth and mortality among species

生长率-死亡率	小树	大树	所有树
RGR_{95}-MR_{25}	0.33 (0.15)	0.59 (0.07)	0.37 (0.10)
RGR_{90}-MR_{50}	**0.49 (0.03)**	0.23 (0.51)	**0.50 (0.02)**
$RGR_{average}$-$MR_{overall}$	0.29 (0.19)	0.02 (0.95)	0.33 (0.14)

注：显著相关显示为粗体。

5.4 讨 论

5.4.1 局域因子对树木死亡与生长影响的异同

在本研究区典型阔叶红松林，生物邻体对所有树木和小树的存活及生长均存在显著的负密度制约效应(图5-1、图5-2)。这说明负密度制约死亡与负密度制约生长效应同时存在于本研究区域。对长白山阔叶红松林的研究发现，在群落水平上，同时存在显著的负密度制约死亡与负密度制约生长效应(Wang et al. 2012, Zhang et al. 2016b)。本章研究结果与长白山研究结果相似，这说明负密度制约生长效应与负密度制约死亡效应均为东北地区温带阔叶红松林物种共存与生物多样性维持的重要机制。

负密度制约效应通常发生在小树阶段或生活史早期阶段，这是由于处于生活史早期阶段的树木与大树相比，通常由于种子扩散限制(seed dispersal limitation)的缘故而定植于次优生境(suboptimal habitats)，导致其对外界环境胁迫(如病虫害、食草动物)的抵抗能力较差；由于其对邻体密度更加敏感，导致资源竞争处于劣势，生长受到削弱，最终容易造成死亡(Russo et al. 2005, Comita et al. 2007)。本章研究结果(图5-1、图5-2)与以往研究结果相似(Wang et al. 2012, Johnson et al. 2014, Zhu et al. 2015b, Wu et al. 2017)。例如，Zhu等(2015b)在巴拿马BCI热带雨林动态监测样地研究发现，同种邻体密度对幼树和中树的存活具有显著的负效应，而对大树的存活具有边缘显著的、微弱的正效应，表明在BCI热带雨林群落中，种内竞争主要发生在树木生活史早期阶段。

此外，随着邻体树种丰富度的提高，以及与邻体之间功能性状差异性的增大，基株的生长率也随之增大(图5-2)，这与第3章发现的系统发育负密度制约死亡相似(图3-2)。被大量功能性状相似性高的邻体所包围的基株往往与邻体之间有较强的资源竞争，因此抑制了基株的生长(Adler et al. 2013, Chen et al. 2016)。本章研究结果与巴拿马BCI热带森林动态监测样地的邻体互补对基株生长存在着正效应，基株的生长率随着与邻体之间功能性状相异性的增大而增大的研究结果类似(Chen et al. 2016)，即为正的邻体互补效应(positive effect of neighborhood complementarity)；此外，地形因子对树木死亡与生长的影响比较相似(图5-1、图5-2)。综上所述，局域因子对树木死亡与生长的影响虽然具有差异性，但差异性并不明显，而相似性较高。

5.4.2 树木死亡与生长的种内及种间关系和物种共存

本章研究表明，在小兴安岭凉水典型阔叶红松林，在种内，树木死亡与生长之间密切相关，尤其是胸径小于35 cm的树木更加明显。当胸径小于35 cm时，树木生长率随胸径急速升高，树木死亡率则随之急速下降。当胸径大于等于35 cm 时，随着树木生长率下降，其死亡率不再急速下降，而是趋于平缓（图 5-1～图 5-3）。使用胸径近似代表树木年龄，可推断出与年龄相关的树木死亡状况。随着树木的衰老，树木的生理功能可能会随之衰退，如光合速率的下降（Lugo and Scatena 1996）。但是近年来也有诸多的研究表明，树木死亡可能并不是因为在生理上的衰老所导致的（Mencuccini et al. 2007, Munné-Bosch 2008, Penuelas and Munné-Bosch 2010, Mencuccini et al. 2014, Munné-Bosch 2015）。树木可能会因遭受外源损伤（例如，木质部机械损伤、昆虫和病原菌的攻击），这些损伤会随着树木年龄的增大而削弱树木的生长活力。Botkin 等（1972）在 JABOWA 森林动态模型树木死亡亚模型中提出，当树木未能维持一个最小的年直径生长量并持续一定时间之后，树木的死亡率将大大增加，甚至树木难以继续存活。Leemans（1991）在 FORSKA 森林动态模型树木死亡亚模型中也应用了相似的方法，他指出树木的死亡率与树木的生长效率（生长活力因子）密切相关，生长活力越小，则树木的死亡概率越高。除了以上机理模型研究以外，经验模型方面对树木死亡的研究也得出了相似的结论。例如，Hamilton（1986）应用 Logistic 回归建模，在美国爱达荷州北部针叶混交林构建了单木枯损模型，发现树木死亡与年直径生长量呈现负相关关系。通常树木在死亡之前具有较低的直径生长率（Pedersen 1998, Bigler and Bugmann 2004）。例如，Cailleret 等（2017）的研究观察到 84%的树木死亡事件中，树木死亡之前，其直径生长率均存在明显的下降趋势。虽然随着树木个体的增大，生长效率通常会随之下降，但是叶面积可能仍然随之增大，这说明整株林木的生物量仍然在增加（Sillett et al. 2010, Sillett et al. 2015）。

生态学中的权衡关系是指在有机体生活史的某一个阶段或某种状态下，当生活史的一个特征的有益变化会涉及对另一特征的不利（或有害）影响时，有机体对适合度"货币"所付出的代价（MacArthur and Levins 1964）。在生活史早期阶段（0～35 cm），树木平均生长率随着径级的增加而上升，然而在生活史末期阶段（>35 cm），树木平均生长率随着径级的增加而下降（图 5-3），这也可以看成是一种树木生长随生活史变异的权衡关系。相似地，在生活史早期阶段（0～35 cm），树木死亡比例随着径级的增加而急速下降，然而在生活史末期阶段（>35 cm），树木死亡比例随着径级的增加不再急速下降，而是趋于平缓（图 5-3），这也可以看成是一种树木死亡随生活史变异的权衡关系，虽然在本章研究结果中这并不明显。为了更合理地分配资源，植物控制着生长、开花、传粉和结实的时机（Bolmgren and Cowan 2008, Kushwaha et al. 2010）。在生活史末期阶段，树木生长通常会随着植物个体的增大或者年龄的增加而下降，这是因为树木积累了呼吸消耗、每单位生物量所持有的叶面积减小、遭受更多的自我遮蔽效应、分配给用于繁殖的资源增多等多重因素所致（Ryan et al. 1997, Mencuccini et al. 2005, Rose et al. 2009）。

随着树木个体增大，树木死亡个体数量出现形似倒"J"形死亡曲线(图 5-3)。这与巴拿马 BCI、马来西亚 Pasoh 等热带雨林动态监测样地研究中随着树木个体的增大，树木死亡率出现"U"形曲线这一结论不同(Wright et al. 2010, Iida et al. 2014a)。本章研究结果与长白山温带森林和八大公山亚热带森林的研究结果类似(Wang et al. 2012, Wu et al. 2017)，在群落水平上，树木的死亡率会随着林木个体增大而下降，林木大小对小树的存活具有显著正效应，但是对大树的存活，效应却不显著，因而并未发现"U"形死亡曲线。

在小树以及所有树中，RGR_{90} 与 MR_{50} 之间存在显著的生长与死亡种间权衡关系，但对于大树则不存在这种显著关系(表 5-1)。植物的生长与死亡种间权衡关系被认为是植物生活史当中最重要的权衡关系之一(另一个重要的权衡关系是繁殖与死亡之间的权衡关系)(Stephenson et al. 2011)。树木死亡率可能会随着树木大小的增加而出现升高、降低或者显示出"U"形曲线，这可能取决于局部环境条件(如光照)中差异性变化，以及植被生长与繁殖之间资源分配(allocation of resources)的转换(Thomas 1996, King et al. 2006, Rüger et al. 2011)。最近热带森林的研究结果表明，树木死亡与生长种间权衡视林木大小而定(size dependent)，本章的研究结果与其相似。Wright 等(2010)在巴拿马 BCI 热带森林动态监测样地的研究发现，有利条件下的快速生长与不利条件下的低死亡率之间存在着显著的种间权衡关系，但这种权衡关系只存在于小树之间，而大树之间不存在这种显著的权衡关系。Iida 等(2014b)在马来西亚 Pasoh 热带森林动态监测样地的研究中同样发现，相对生长率与死亡率之间存在显著的正相关关系，即权衡关系，但这种权衡关系也只存在于小树阶段。

树木死亡与生长之间存在显著的种间权衡关系(表 5-1)。Kitajima 和 Poorter (2008)认为树木死亡-生长种间权衡关系能够促进物种共存。这种促进物种共存的权衡关系可能是由于生态位分化及适合度趋同共同造成的。森林群落中的 r 对策者，例如，某些先锋树种，其多喜光，生长速度快但寿命短，具有较高的死亡率。r 对策树种将资源主要分配于生长能力，而这种取舍导致投资于竞争能力及存活能力的资源不足。相反，某些树种属于 K 对策者，例如，某些耐阴树种，其生长缓慢但寿命长，具有较高的存活率。K 对策树种将资源主要投资于竞争及存活能力，而导致投入于快速生长能力的资源不足。实际上，林隙更新是典型阔叶红松林群落的演替特征(李景文 1997)。当红松死亡后，阔叶树种，特别是杨、桦等具有较高生长率的喜光、先锋树种，首先占据红松死亡后所形成的林隙位置，使得红松幼树被压制而处于林下层，但是幼年红松具有耐阴性且寿命长，当林隙发育到一定时期之后，红松逐渐进入并占据林冠层，杨、桦等喜光树种因此逐渐被压制而导致死亡。落叶树种和红松在一个林隙斑块内发生着相互交替取代对方的周期性动态过程(Ge 1994, Ge et al. 1995)，并维持典型阔叶红松林群落生物多样性。

r 对策喜光树种通常只有占据林冠层以确保得到足够的光照才能存活,而某些 K 对策耐阴树种则在庇荫的林下层能够存活,不同生态对策的树种对光环境的适应可能有所差异。竞争排除原理认为,具有完全相同资源需求的两个物种不能长期稳定共存,必然导致其中一个物种排除掉另一个物种,而不同生态对策的树种占据不同的生态位,资源生态位分化可有利于促进物种共存并维持森林群落生物多样性。另外,权衡轴上位置相同或相近的树种之间具有相同或相似的适合度,根据当代物种共存理论,适合度趋同有助于促进物种共存。例如,先锋树种通常具有相似的适合度(存活率和生长率),它们的竞争能力相似,这样先锋树种内部任何一个物种都不会轻易地通过竞争排除将其他物种排除掉,从而促进物种共存。根据当代物种共存理论,有两种情况可促进物种共存,即物种之间的生态位差异越大、平均适合度差异越小(Chesson 2000)。喜光树种与耐阴树种可能通过资源生态位分化实现稳定共存,而在喜光树种或耐阴树种内部可能通过适合度趋同而实现稳定共存。种间生态位分化与适合度趋同共同作用,最终促进整个森林群落物种共存,因此推断,稳定化机制与均等化机制可能是凉水典型阔叶红松林物种共存与生物多样性维持的重要机制。

5.5 本章小结

本章探索了树木死亡-生长种间权衡与物种共存之间的关系,并借助广义线性混合模型等手段,揭示了局域因子对树木死亡与生长影响的差异。结果表明,研究区树木死亡与生长之间存在显著的种间权衡关系,特别是对于小径级树木,死亡率越高的树种,其生长率也越高,反之亦然,这与热带森林的研究结果基本一致,可能导致种间生态位分化与适合度趋同共同发生作用,进而促进物种共存并维持森林群落生物多样性。局域因子对树木死亡与生长的影响存在差异,但差异并不明显。同时存在负密度制约死亡与生长效应,小径级树木受这种影响更为明显。地形因子对树木死亡与生长的影响比较相似。稳定化机制与均等化机制可能是凉水典型阔叶红松林物种共存与生物多样性维持的重要机制。

6 树木死亡–更新种间权衡与物种共存

6.1 引 言

负密度制约与生境异质性通过对树木死亡的影响促进物种共存；通过树木死亡–生长种间权衡也可对促进物种共存具有一定贡献。死亡是更新的开始。与树木生长和死亡类似，树木更新也属于大型森林动态监测样地中的一个重要的调查项目(Anderson-Teixeira et al. 2015)。以往研究表明，木本植物死亡与更新分别对森林群落物种共存和生物多样性维持具有重要影响(Harms et al. 2000, Zhu et al. 2015b)。

与树木死亡和生长相似，树木更新也受生物与非生物因子综合作用。同种或系统发育负密度制约效应不仅对同种或亲缘种基株树木和幼苗的存活与生长起到抑制作用，而且对同种或亲缘种基株树木的更新也同样产生不利影响，这对物种共存及维持群落生物多样性有利(Harms et al. 2000, Zhu et al. 2015a, Zhu et al. 2015b)。此外，树木更新也受到光照等非生物因子的影响(Adams et al. 2013)。再者，以往诸多研究都是只关注局部环境因子对某一项森林动态(如生长、存活或更新)的影响(Harms et al. 2000, Zhu et al. 2015b, Chen et al. 2016)，并深入探讨森林群落物种共存机制，然而，在典型阔叶红松林，局域因子对树木死亡与更新的影响是否存在差异目前尚不清晰。

以往研究发现树木死亡与生长之间往往存在显著的种间权衡关系(Wright et al. 2010, Iida et al. 2014b)。树木死亡与生长种间权衡关系促进物种共存并维持生物多样性(Iida et al. 2014b)。以往关于树木死亡与更新关系的研究多集中于物种死亡率和更新率等统计计算，并分析死亡率与更新率差异(汪殷华等 2011, 葛结林等 2012, 金毅等 2015, 王慧杰等 2016)。然而，在典型阔叶红松林，局域因子对树木死亡与更新的影响是否存在差异，树木死亡与更新之间是否如同树木死亡与生长一样也存在种间权衡关系，进而促进物种共存，目前尚不明确。

在本章中，使用2010~2015年共5年间的树木死亡与更新动态监测数据，探究树木死亡–更新种间权衡与物种共存机制(稳定化机制与均等化机制)之间存在何种关系，并试图回答以下两个问题：①局域因子对树木死亡与更新的影响是否存在差异？在样方尺度下，是否同时存在负密度制约死亡与更新效应？②探索树木死亡与更新之间存在何种种间关系？是否进而促进物种共存？

相对应的提出两点假设：①局域因子对树木死亡与更新的影响存在差异，同时存在负密度制约死亡与更新效应；②树木死亡与更新之间存在显著的种间权衡关系，进而促进物种共存。

6.2 研究方法

6.2.1 数据收集

与树木死亡动态的缓慢性不同，树木的更新进程往往快于树木死亡。为了尽可能研究具有较高完整度生活史阶段的树木死亡和更新动态，本章选取 2010~2015 年共 5 年间的树木死亡和更新动态监测数据集。将 5 年间新增的 DBH≥1 cm 的活的木本植物视为补员（recruit）。非生物因子为 5 m、10 m 及 20 m 空间尺度下的地形因子（包括海拔、凹凸度和山体阴影）。

6.2.2 样方尺度树木死亡与更新驱动因子的构建

本章模型自变量包括：样方内初始树木平均胸径（AvgDBH）、净亲缘关系指数（NRI），以及海拔、凹凸度和山体阴影 3 个地形因子。由于第 3 章中已经根据 AIC 模型拟合优度判断准则确定了胸径、生物邻体与地形因子为最优变量组合，同时为了与第 3 章模型生境因子保持一致性，本章仅将地形因子视为非生物因子。

6.2.3 样方尺度树木死亡与更新模型

树木死亡模型中因变量为每一个样方内 2010~2015 年共 5 年间的树木死亡的株数。树木更新模型中因变量为每一个样方内 5 年间的树木更新的株数。树木更新模型通常在样方尺度下建立（Rüger et al. 2009, Zhang et al. 2012b, Zhu et al. 2015a）。树木死亡模型与更新模型的因变量属于计数变量，因此广义线性模型中采用 Poisson 回归分析。负二项回归模型、零膨胀 Poisson 回归模型及零膨胀负二项回归模型是对 Poisson 回归模型的完善和补充，近年来上述计数型回归模型也逐渐应用于树木更新模型研究中（Rüger et al. 2009, Zhang et al. 2012b）。因此本章采用 Poisson 回归模型、负二项回归模型、零膨胀 Poisson 回归模型及零膨胀负二项回归模型分别在 3 个样方尺度（5 m、10 m 和 20 m）下建立树木死亡和更新模型。采用标准差标准化法（Z score）对所有自变量进行无量纲化处理，以便允许各个模型参数之间可以直接地比较大小。关于以上模型详情参见 4.2.3 节。

负二项回归模型采用 R 软件 MASS(Zeileis et al. 2008)程序包拟合。零膨胀 Poisson 回归模型及零膨胀负二项回归模型拟合、Vuong 检验采用 R 软件 pscl (Zeileis et al. 2008)程序包实现。

6.2.4 树木死亡与更新种间关系检验

各个树种死亡率(M)与更新率(R)计算公式如下：

$$M = (\ln N_0 - \ln S_t)/T \tag{6-1}$$

$$R = (\ln N_t - \ln S_t)/T \tag{6-2}$$

式中，N_0 代表某树种在 2010 年调查时的个体数；N_t 是该树种在 2015 年调查时的个体数；S_t 是该树种在 2015 年调查时的存活个体数；T 为两次调查时间间隔，即 5 年。

对多度至少达到 100 的树种计算其死亡率和更新率。采用 Pearson 相关性分析检验树木死亡与更新种间权衡关系。

6.3 结　　果

6.3.1 局域因子对树木死亡与更新的影响

2010~2015 年 DBH<5 cm 的死亡木及全体死亡木的空间分布规律并不明显，然而 DBH≥5 cm 的死亡木分布特征较为明显：样地海拔较低的西南区域内死亡木密度明显大于其他地区(图 6-1)。2010~2015 年共 5 年间的更新木分布特性比较明显：在样地海拔较高且陡峭处密集分布，而在海拔较低且平坦地区更新木密度较低(图 6-1)。

在 5 m 及 10 m 样方尺度下、任一类型的计数型模型下，净亲缘关系指数对树木更新均存在显著的负效应，而在 20 m 样方尺度下，这种负效应则不显著(图 6-2)。无论在任一样方尺度、任一类型的计数型模型下，海拔对树木死亡均存在显著的负效应；相反，海拔对树木更新均存在显著的正效应(图 6-2)。无论在任一样方尺度、任一类型的计数型模型下，山体阴影对树木更新均存在显著的正效应，然而对树木死亡则无显著影响(图 6-2)。

6.3.2 树木死亡与更新的种间关系

根据各个树种死亡率和更新率的 Pearson 相关系数探索可知,死亡率与更新率之间存在显著的种间正相关($r=0.69$,$P<0.001$),即 2010~2015 年共 5 年间死亡率越高的树种,其更新率也越高;死亡率越低的树种,其更新率也越低。

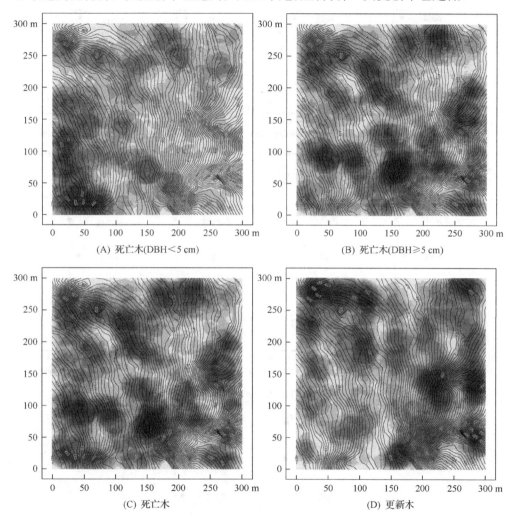

图 6-1 2010~2015 年死亡木与更新木空间点密度分布(深色代表密度较高,浅色代表密度较低)

Fig. 6-1 Point density plot showing the distribution of dead trees and recruits from 2010 to 2015 (Dark and light represent higher and lower densities for trees, respectively)

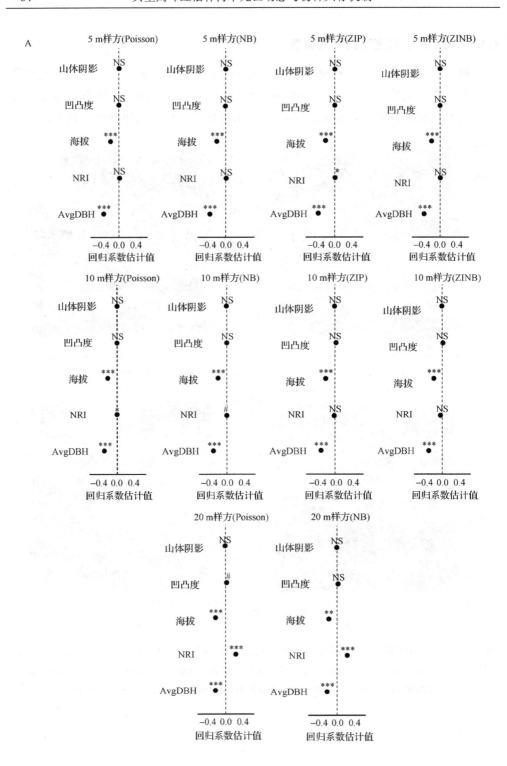

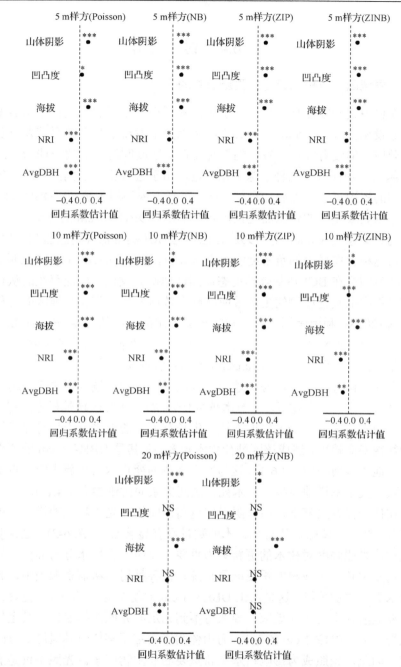

图 6-2 2010~2015 年样方尺度下局域因子对树木死亡(A)与更新(B)的影响(根据 Vuong 检验,负二项回归为最优模型形式)

Fig. 6-2 Standardized parameter estimates (± SE) of local variables (AvgDBH, NRI and topography) on tree mortality and recruitment from 2010 to 2015 at different spatial scales. Negative binomial regression is the best structure via Vuong's test

6.4 讨 论

6.4.1 局域因子对树木死亡与更新影响的异同

在典型阔叶红松林，局域因子对树木死亡与更新的影响存在明显差异，这与研究假设相符。在 5 m 和 10 m 样方尺度下，发现了显著的系统发育负密度制约更新效应，但没有发现一致性的系统发育负密度制约死亡效应(图 6-2)。换言之，NRI 越大，则样方内物种亲缘关系越近，由于近缘物种之间的资源竞争比远缘物种之间的资源竞争强度更加激烈，抑制了样方内的树木更新，DBH<1 cm 的幼苗由于难以获取充足的资源而死亡，无法晋级为 DBH≥1 cm 的补员，导致样方内更新木的数量越低。Condit 等(1992)在巴拿马 BCI 热带森林动态监测样地研究发现，少数常见种存在显著的负密度制约更新效应，符合 Janzen-Connell 假说预测。Harm 等(2000)同样在 BCI 热带森林动态监测样地研究发现，广泛存在的从种子发育到幼苗过程中的负密度制约更新效应能够有效地增加热带森林地区的物种多样性。Zhu 等(2015a)利用美国农业部林务局美国东部地区的 FIA(Forest Inventory and Analysis)森林调查数据分析发现，该地区广泛存在同种邻体负密度制约更新效应，特别是常见种，研究结果与 Janzen-Connell 假说相一致。

在任一样方尺度下，海拔越高的区域，其死亡木越少，更新木越多；海拔越低的区域，其死亡木越多，更新木越少(图 6-2)。西南区域的样方海拔较低、地势平坦，由于该区域曾受过轻度干扰，生长有大量的枫桦、山杨等先锋树种，这些树种往往寿命较短，造成该区域死亡木较多，尤其是 DBH≥5 cm 的死亡木聚集分布在样地的西南区域(图 6-1)，这与实际观测的死亡木与更新木空间点密度分布显示一致。关于本样地海拔与树木死亡之间关系的讨论参见 4.4.1 节。

在任一样方尺度下，山体阴影值越高，即相对光照越强的样方，更新木的数量越多，然而山体阴影值对死亡木的数量没有显著影响(图 6-2)。这说明相比于树木死亡，光照强度对树木的更新更为重要。光照、土壤水分与养分是植物进行光合作用及生长不可或缺的重要能量来源。在各种树木所需资源当中，光照可能对于树木更新更加关键。这是由于 DBH<1 cm 的幼苗处于林下层，更容易受到林冠层、亚冠层林木的遮光效应，导致得不到充足的光照而抑制生长甚至导致死亡，无法晋级为 DBH≥1 cm 的补员，而山体阴影决定了某样方林冠层上方的相对光照强度。因此，光照成为影响树木更新重要的非生物因子。光照不仅是影响温带森林树木生长至关重要的资源(McMahon et al. 2011)，而且 Rüger 等(2009)在巴拿马 BCI 森林动态监测样地的研究还发现，几乎所有树种的更新均与光照有效性呈正相关关系，换言之，光照越强则对树木更新的促进作用越强，表明光照有效性的时空变异有利于维持热带森林生物多样性。山体阴影可通过对光强的再分配，间

接地对树木更新产生影响。

综上所述，局域因子对树木死亡与更新的影响具有明显的差异性，这与第 5 章研究得到的局域因子对树木死亡与生长的影响差异性不明显的结论有所不同。

6.4.2 树木死亡–更新种间权衡与物种共存

本研究中死亡率与更新率之间存在显著的种间正相关（$r = 0.69$，$P<0.001$），即树木死亡与更新之间存在种间权衡关系，这与研究假设相符合。死亡率越高的树种，其更新率也越高；死亡率越低的树种，其更新率也越低。

中性理论假设群落中所有物种具有相等的出生率和死亡率（Tilman 2004），然而这在现实森林群落中很少被观察到。Zhang 等（2012a）的研究突出强调了统计权衡（demographic trade-offs），如出生–死亡率权衡（birth-death trade-off），对于促进森林群落物种共存及维持生物多样性具有重要意义。与树木死亡–生长种间权衡关系类似，r 对策树种繁殖率高，但投资于竞争能力及存活能力的资源不足，而 K 对策树种恰好与 r 对策树种相反，因此出现死亡–繁殖种间权衡关系。

以往研究证实负密度制约死亡、负密度制约更新、树木死亡–生长种间权衡及出生–死亡率权衡可促进森林群落物种共存并维持生物多样性（Janzen 1970, Connell 1971, Harms et al. 2000, Kitajima and Poorter 2008），那么树木死亡–更新间权衡是否也同样促进物种共存？目前尚无充分证据排除这种可能性。本章研究结果与第 5 章研究结果类似，树木死亡–更新种间权衡关系也可能是生态位分化及适合度趋同共同造成的。r 对策喜光树种通常只有占据林冠层以确保得到足够的光照才能存活，而某些 K 对策耐阴树种则在庇荫的林下层也能存活。不同生态对策的树种占据不同的生态位，资源生态位分化有利于促进物种共存并维持森林群落生物多样性。树种出生率、繁殖率与更新率虽存在差异，但三者间关系紧密。出生率、繁殖率越高的树种，则其更新率可能也越高，反之亦然。样地内调查树种出生率、繁殖率的难度较大，而更新率相对容易进行调查，更新调查同时也是森林动态样地复查时三个主要复查项目之一（Anderson-Teixeira et al. 2015）。另外，权衡轴上位置相同或相近的树种之间具有相同或相似的适合度，根据当代物种共存理论，适合度趋同有助于促进物种共存（Chesson 2000）。例如，先锋树种可能因具有相似的适合度（存活率和更新率），致使其竞争能力相似，因此先锋树种内部任何一个物种都不会轻易地通过竞争排除将其他物种排除掉，从而促进物种共存。喜光树种与耐阴树种可能通过资源生态位分化实现稳定共存，而在喜光树种内部或耐阴树种内部可能通过适合度趋同而实现稳定共存。种间生态位分化与适合度趋同共同作用，最终促进整个森林群落物种共存。

本章探索了树木死亡–更新种间权衡关系，未来应进一步完善并更加深入地探究树木死亡–更新种间权衡，以及其他种间统计权衡与物种共存之间的关系及其在温带与热带森林中的普适性。

另外，第3章、第4章、第5章、第6章模型的自变量设置具有一定的局限性，由于在个体尺度及样方尺度下分别建立模型，难以保证所有模型的自变量完全一致，在以后的研究中应注意到这种情况。

6.5 本章小结

本章在物种共存理论研究中探索了温带森林树木死亡-更新权衡与物种共存之间的关系，并运用Poisson回归、负二项回归、零膨胀回归模型等手段，揭示了局域因子对树木死亡与更新影响的差异。结果表明，研究区树木死亡与更新之间存在显著的种间权衡关系，死亡率越高的树种，其更新率也越高；反之亦然。这种权衡所导致的种间生态位分化与适合度趋同可促进物种稳定共存。研究区树木死亡-更新种间权衡有可能促进物种共存，但该结论在其他温带森林及热带森林是否具有普适性，仍然需要未来更深入的研究。局域因子对树木死亡与更新的影响存在差异，并且差异比较明显。在5 m和10 m样方尺度下，可检验出显著的负密度制约更新效应，但没有发现一致的负密度制约死亡效应。在5 m、10 m或20 m样方尺度条件下，海拔越高的区域，其死亡木越少，更新木越多；反之亦然。

7 物种共存理论修正树木死亡过程模型

7.1 引　　言

表征林木死亡格局对于理解森林动态至关重要，明确死亡树木的时空分布更有助于理解树木死亡的特征，是进一步研究树木死亡规律及物种共存的基础条件。

稳健的树木死亡算法是植被动态模型（dynamic vegetation model, DVM）中重要的组成部分之一（Manusch et al. 2012）。树木死亡模型可以分为两类：一类是基于调查数据的经验模型建模方式；另一类是基于过程的机理模型建模方式（Hülsmann et al. 2017）。基于过程的树木死亡模型明确地集成了死亡机理，并且具有更强的综合性；而树木死亡的经验模型通常具有更好的简洁性，并且暗含树木死亡机理（Adams et al. 2013）。此外，基于过程的树木死亡建模方式通常是一个树木生长效率（生长受压死亡）和林木大小（内禀死亡）的一个函数（Manusch et al. 2012）。树木死亡经验模型一般基于统计回归的 Logistic 回归，近年来多用于森林经营管理，以及探讨物种共存与森林群落构建机制（Mencuccini et al. 2005, Grayson et al. 2017, Wu et al. 2017, Zhang et al. 2017）。根据关注点的不同，树木死亡经验模型仍可进一步细分为预测模型和解释模型。但是，树木死亡经验模型中所采用的时间分辨率均依赖于调查间隔，难以预测和观察微尺度下的树木死亡时空动态。树木死亡过程模型能够弥补经验模型的上述缺陷。但是，以往的树木死亡过程模型往往缺失树木的定位坐标、生境异质性信息和树木死亡动态监测数据，造成模型评价难度大，模型检验只能进行确证性检验而无法进行有效性检验，不得不从裸地开始进行模拟（Botkin et al. 1972, Busing 1991, Leemans 1991, 桑卫国和李景文 1998, Yan and Shugart 2005, 国庆喜和葛剑平 2007）。

近年来，随着大型森林动态监测样地的出现和发展，为树木死亡过程模型的发展提供了新的契机（Harms et al. 2000, Uriarte et al. 2004, Wright et al. 2010, Wang et al. 2012）。大型固定样地不仅提供了树木坐标，而且提供了树木死亡动态监测数据，为树木死亡过程模型的研制和模型精度评价提供了基础数据。虽然近年来，有不少学者依托大型森林动态监测样地，采用树木死亡经验模型研究了物种共存和群落构建机制（Johnson et al. 2014, Wu et al. 2017），但是，其少有将大型森林动态监测样地和树木死亡过程模型集成起来的报道。

此外，与大型森林动态监测样地一同发展起来的物种共存理论的最新研究成果(如邻体半径种间变异、生境异质性、负密度制约效应)可为经典树木死亡过程模型的部分结构的改进和修正提供借鉴,对原模型细节之处进行进一步刻画。例如,前文研究结果表明,负密度制约、生境异质性可通过对树木死亡的影响进而促进物种共存;种间生态位分化与适合度趋同通过树木死亡-生长种间权衡、树木死亡-更新种间权衡也可能促进物种共存。

本章在凉水典型阔叶红松林 9 hm² 大型森林动态监测样地的支持下,借鉴物种共存理论研究成果,修正经典树木死亡过程模型的部分结构,并利用树木死亡动态监测数据进行模型精度评价。本章研究是对树木死亡过程模型与森林动态样地集成的一次尝试,以期为以后的相关研究提供新的思路和角度。

7.2 研究方法

7.2.1 模型结构与计算公式

根据用于描述基于个体/主体的模型(individual-based and agent-based model, IBM and ABM)的标准规范,即 ODD(overview, design concept, detail)规范(Grimm et al. 2006, Grimm et al. 2010),描述本研究研制的树木死亡动态模拟模型。

7.2.1.1 模拟目的

树木死亡动态模拟(tree mortality dynamic simulator, TMDS)模型是在充分借鉴经典林窗模型(如 JABOWA、FORSKA、SPACE)中树木死亡模块结构和算法的基础上,结合森林群落物种共存理论研究成果进一步修正和改进后研制而成。树木死亡动态模拟模型主要用于模拟微尺度下森林群落树木死亡的时空动态。

7.2.1.2 模型实体、状态变量(和/或属性)及模拟尺度

实体主要包括树木(ESRI Shapefile,矢量数据格式)和样方(GeoTIFF,栅格数据格式)。树木的状态变量与属性包括树号、树种、坐标、胸径、树高等。样方的状态变量与属性包括山体阴影、土壤湿度。以 1 年为时间步长,以 5 m 为空间分辨率。

为了尽可能研究较长时间的树木死亡动态,本章选取 2005~2015 年共 10 年间的树木死亡动态监测数据。在 2005 年,对样地内所有的 DBH≥2 cm 的活的木本植物进行编号挂牌、测量胸径、定位坐标并鉴别树种。2015 年,对 2005 年调查的活立木进行复查,记录其存活状态(存活或死亡)。地形因子包括海拔、凹凸度和山体阴影。

本研究采用凉水样地 19 个树种作为研究对象，分别为红松、红皮云杉、鱼鳞云杉、臭冷杉、紫椴、糠椴、水曲柳、枫桦、白桦、春榆(*Ulmus japonica*)、裂叶榆、胡桃楸、黄菠萝、蒙古栎、山杨、大青杨、色木槭、青楷槭和花楷槭。对于树种参数，本研究借鉴并采纳了前人的研究成果。邵国凡提供了阔叶红松林 8 个主要树种的干燥度指数和最大、最小积温参数(邵国凡 1991)；葛剑平提供了小兴安岭典型阔叶红松林 5 个主要树种的生长效率阈值参数(葛剑平 1996)；桑卫国和李景文提供了小兴安岭凉水典型阔叶红松林 18 个主要树种的参数(桑卫国和李景文 1998)。对于某些前人研究没有提供的树种参数，本研究借助系统发育树及种间亲缘关系距离近似求取相关树种参数(表 7-1～表 7-3)。模型中的环境参数主要参考了桑卫国等(1999)、桑卫国和李景文(1998)在小兴安岭典型阔叶红松林开展的研究成果。

表 7-1　树木死亡模型参数
Table 7-1　Explanation of tree mortality model parameters

参数名称	符号	单位
立地环境参数		
生长季平均入射光强	I_0	$\mu mol/(m^2 \cdot s)$
群落消光系数	k	
有效积温	DEGD	$(℃ \cdot d)/a$
干燥度	DI	
树种参数		
光半饱和点	α	$\mu mol/(m \cdot s)$
光补偿点	c	$\mu mol/(m \cdot s)$
总光合速率	γ	$cm^2/(m \cdot a)$
边材维持成本	δ	$cm^2/(m^2 \cdot a)$
内禀死亡率	U_0	
生长受压死亡率	U_1	
风倒木死亡率	U_2	
最小积温	$DEGD_{min}$	$(℃ \cdot d)/a$
最大积温	$DEGD_{max}$	$(℃ \cdot d)/a$
最大干燥度	DI_{max}	
生长效率阈值	θ	
邻体半径	R	m
最大胸径	DBH_{max}	cm

表 7-2 立地环境参数表
Table 7-2 Site parameters

符号	生长季平均入射光强	群落消光系数	有效积温	干燥度
参数值	450	0.4	1700	0.74

注：具体参数解释查看表 7-1。

表 7-3 树种参数表
Table 7-3 Species parameters

树种	c	α	γ	δ	DBH_{max}	U_0	U_1	$DEGD_{min}$	$DEGD_{max}$	DI_{max}	θ
红松 Pinus koraiensis	14.78	250	14.56	0.045	200	0.009	0.04	1350	3250	1.05	0.025
红皮云杉 Picea koraiensis	5.69	175	5.81	0.016	100	0.015	0.06	800	1800	0.89	0.025
鱼鳞云杉 Picea jezoensis	5.69	175	5.81	0.016	100	0.015	0.06	800	1800	0.89	0.025
臭冷杉 Abies nephrolepis	8.53	180	9.11	0.024	60	0.023	0.06	650	1750	0.96	0.025
紫椴 Tilia amurensis	12.09	260	9.08	0.024	100	0.015	0.06	1500	3600	1.4	0.025
糠椴 Tilia mandshurica	12	250	12.2	0.026	100	0.015	0.06	1500	3600	1.4	0.025
水曲柳 Fraxinus mandschurica	24	400	31	0.094	100	0.018	0.06	1600	3400	1.25	0.025
枫桦 Betula costata	30	420	50.24	0.131	82	0.023	0.12	1350	4000	1.7	0.03
白桦 Betula platyphylla	50	650	78.39	0.136	60	0.046	0.12	1350	4000	1.7	0.04
春榆 Ulmus japonica	25.4	300	21.06	0.055	90	0.023	0.06	1250	3600	1.55	0.025
裂叶榆 Ulmus laciniata	24	300	21.78	0.051	70	0.023	0.06	1250	3600	1.55	0.025
胡桃楸 Juglans mandshurica	20	400	22.07	0.038	80	0.018	0.06	1600	3800	1.32	0.025
黄檗 Phellodendron amurense	19	400	19.81	0.038	100	0.018	0.06	1650	3670	1.31	0.025
蒙古栎 Quercus mongolica	41.4	600	133.7	0.324	60	0.013	0.06	1450	3800	1.64	0.025
山杨 Populus davidiana	37.5	550	30.41	0.066	60	0.031	0.12	1400	4000	1.75	0.04
大青杨 Populus ussuriensis	36	550	50.04	0.13	150	0.023	0.12	1250	3600	1.75	0.04
色木槭 Acer mono	28	300	15.71	0.027	60	0.031	0.06	1450	3800	1.39	0.025
青楷槭 Acer tegmentosum	28	300	15.71	0.027	30	0.031	0.06	1450	3800	1.39	0.025
花楷槭 Acer ukurunduense	28	300	15.71	0.027	30	0.031	0.06	1450	3800	1.39	0.025

注：具体参数解释查看表 7-1。

7.2.1.3 模型主要过程概览及模拟流程

树木死亡过程主要包括树木内禀死亡、生长受压死亡（包含树木生长过程）与外部干扰死亡。生长受压死亡过程包含树木生长过程。树木生长过程初步涉及光合作用、呼吸作用等生理生态过程。异速生长方程用于计算生物量和叶面积。模型流程图见图 7-1。

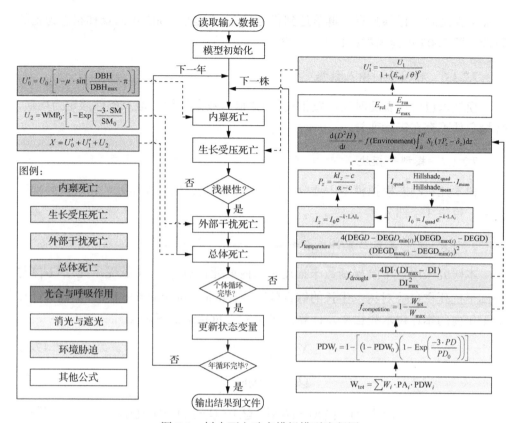

图 7-1 树木死亡动态模拟模型流程图

Fig. 7-1 Flowchart of tree mortality dynamic simulator

7.2.1.4 模型设计概念

适应策略(adaptation)——树木高生长与胸径生长策略固定不变。相互作用(interaction)——基株死亡与生长过程直接受其他邻体影响，邻近树木之间存在光等资源竞争关系，包括种内和种间竞争。敏感性(sensing)——树木生长过程强烈地受到光环境影响，而基株的光环境主要受到地形(山体阴影)及邻体树冠遮挡影响。树木死亡中的生长受压死亡过程直接受到树木生长过程影响。随机性(stochasticity)——树木死亡为随机过程，重复运行模拟程序 50 次，求取平均预测死亡率。

7.2.1.5 模型初始化

实体(包括树木和样方)的所有状态变量和属性均为非随机选取，而是来自于森林动态监测样地实测的动态监测数据。与以往多数林窗模型研究不同，该模型并非从裸地起进行预热直至群落达到稳定状态,而是以 2005 年的森林状态为初始阶段,即 2005 年为模型模拟的第一年。

7.2.1.6 时间序列输入数据

模型未包含代表随时间而发生变化的"输入数据"。模拟地点条件假定为均质,环境条件不存在年际变异。

7.2.1.7 模型子模块

树木死亡动态模拟模型主要分为内禀死亡模块、生长受压死亡模块及外部干扰死亡模块 3 个部分。

7.2.1.8 死亡函数与累计死亡概率

本研究采用一个阶梯函数来描述树木内禀死亡(U_0')、生长受压死亡(U_1')与外部干扰死亡(U_2)对树木生长的综合效应(Leemans and Prentice 1987)。同时计算若干年内的累计树木死亡率。计算公式参考了 FORSKA 模型。

$$X = U_0' + U_1' + U_2 \tag{7-1}$$

$$Q = 1 - \mathrm{Exp}(-X) \tag{7-2}$$

$$P_n = 1 - (1-Q_1)(1-Q_2)(1-Q_3)\cdots(1-Q_i)\cdots(1-Q_n) \tag{7-3}$$

式中,X 代表树木死亡率;Q 代表树木年度死亡率;Q_i 代表树木第 i 年的死亡率;P_n 代表 n 年内的累计死亡率。

(1)内禀死亡:以往模型将每个特定树种的内禀死亡率视为一个随时间不变的常数,这使得树木年龄越大,其累计的树木死亡率越高,产生了一条负指数存活曲线,而这与实际情况并不相符(Bugmann et al. 1996)。通常情况下,老树和幼树的内禀死亡率往往高于成年中龄树。此外,Manusch 等(2012)建议在模拟树木内禀死亡率时,宜采用树木胸径代替树木年龄。因此,本研究建立的模型采用了一个正弦函数(梁玉莲 2010),并用胸径代替年龄对内禀死亡率函数进行动态校正:

$$U_0' = U_0 \cdot \left[1 - \mu \cdot \sin\left(\frac{\mathrm{DBH}}{\mathrm{DBH}_{\max}} \cdot \pi\right)\right] \tag{7-4}$$

式中,U_0 代表依赖年龄的内禀死亡率;DBH 代表树木当前胸径;$\mathrm{DBH}_{\max}$ 代表特定树种的最大胸径;μ 代表校正系数,设置为 0.25。由于即使成年中龄树也应具有一定的内禀死亡率，因此模型设置校正系数，以防止内禀死亡率出现零值。

(2) 生长受压死亡：模型同样采用了依赖生长的死亡率，即生长受压死亡率。树木的生长率和寿命之间往往存在权衡关系，即生长较慢的树木(如演替末期树种)的寿命一般比生长较快的树木(如先锋树种)更长(Bigler and Veblen 2009, Rötheli et al. 2011)。第 5 章研究发现，典型阔叶红松林存在树木死亡-生长权衡种间关系(图 5-4)，并且这种权衡促进物种共存并维持森林生物多样性，它由树种生态对策差异及资源生态位分化原因导致。然而，以往大多数研究将所有树种的生长受压死亡率视为一个定值。因此，本研究对此进行修正，TMDS 模型假设生长较快的先锋树种树木(如桦木属和杨属)20 年后只有 10%的树木存活，每年生长受压死亡率为 0.12。而对于生长较慢的演替末期树种树木(如松属)，则假设 60 年后只有 10%的树木存活，每年生长受压死亡率为 0.04。对于其他树种，则假设 40 年后只有 10%的树木存活，每年生长受压死亡率为 0.06。模型中使用生长效率来判定树木是否处于生长受压状态，生长受压与生长效率相关。

当树木个体生长效率连续 SurvYrs 年低于生长效率阈值 θ 时，这株树存活的概率为 SurvProb%，所以得到树木生长衰弱的死亡概率为

$$U_1 = 1 - \mathrm{Exp}\left[\frac{\ln(\mathrm{SurvProb})}{\mathrm{SurvYrs}}\right] \tag{7-5}$$

$$U_1' = \frac{U_1}{1+(E_{\mathrm{rel}}/\theta)^\rho} \tag{7-6}$$

式中，U_1' 代表修正后的树木生长受压死亡率；E_{rel} 为生长效率；ρ 为陡度参数。

7.2.1.9 生长效率

FORSKA 模型在计算树木依赖生长的死亡率时，将树木的生长效率作为胁迫因子校正生长受压死亡率。生长效率定义为树木实际生长率和最大生长率之间的比值(Leemans and Prentice 1987)。当树木的年生长量长期低于某一特定数值时，树木的死亡率会急剧升高。Leemans 和 Prentice (1987)将特定树种的这一特定数值定义为生长效率阈值。本研究在模型中也采用了生长效率的概念。

$$P' = \frac{\int_B^H P_z \mathrm{d}z}{H-B} \tag{7-7}$$

$$H' = \frac{H+B}{2} \tag{7-8}$$

$$E_{\text{rel}} = \frac{E_{\text{rea}}}{E_{\text{max}}} = \frac{f(\text{Environment})(\gamma P' - \delta H')L}{\gamma P_0 L} = f(\text{Environment})\left(\frac{P'}{P_0} - \frac{\delta H'}{\gamma P_0}\right) \quad (7\text{-}9)$$

式中，E_{rel} 代表生长效率；E_{rea} 代表树木实际生长率；E_{max} 代表树木最大生长率；P_0 代表在没有遮挡条件下树冠平均光合作用响应值。

树木死亡动态模拟(TMDS)主要借鉴了 FORSKA 模型和 SPACE 模型中的核心算法。以往模型大多根据生态环境的综合作用规律建立环境函数，即认为环境是由许多因子组成的综合体，各因子之间是相互独立的，并且分别作用于树木生长，如 JABOWA、FORSKA 和 SPACE 模型。而采用这种方式的主要问题是人为地产生了较低的环境综合影响值，放大了环境因子间作用的部分，使得环境胁迫对树木生长的作用过大。采用 Liebig 最小因子限制定律(Liebig's law of the minimum)能够合理地解决这一问题(Bugmann 2001)。TMDS 模型中计算环境胁迫对树木生长的综合作用因子时采用了不同于以上模型的 Liebig 最小因子限制定律，即树木生长主要受最优环境差距最大的因子来控制(Foster et al. 2017)。

$$f(\text{Environment}) = \min(f_{\text{temperature}}, f_{\text{drought}}, f_{\text{competition}}) \quad (7\text{-}10)$$

式中，$f_{\text{temperature}}$、f_{drought} 和 $f_{\text{competition}}$ 分别代表温度、湿度和竞争限制因子对树木生长的胁迫作用。

7.2.1.10 温度限制因子

温度限制因子即热量指数。构建温度限制因子来表达温度对每一个树种生长的胁迫作用(Botkin et al. 1972，邵国凡 1991)。采用以 10℃为基准计算有效积温(effective growing degree-days, DEGD)作为温度对树木生长的影响。温度限制函数是一个抛物线函数，在积温的最大值和最小值处，该函数值为 0。

$$f_{\text{temperature}} = \frac{4(\text{DEGD} - \text{DEGD}_{\min(i)})(\text{DEGD}_{\max(i)} - \text{DEGD})}{(\text{DEGD}_{\max(i)} - \text{DEGD}_{\min(i)})^2} \quad (7\text{-}11)$$

式中，$f_{\text{temperature}}$ 代表积温对树木生长的影响；DEGD 代表研究区年积温；$\text{DEGD}_{\max(i)}$ 和 $\text{DEGD}_{\min(i)}$ 分别代表树种 i 能忍耐积温的两个极值(最大积温和最小积温)，两个极值可根据树种 i 地理分布区的南北界限并根据积温分布图来得到(Botkin et al. 1972)。

7.2.1.11 湿度限制因子

湿度限制因子即水分指数。采用干燥度来表达湿度对每一个树种生长的胁迫作用，干燥度定义为潜在蒸散(Thornthwaite and Mather 1957)与年降水量之比。本研究引用邵国凡(1991)的研究构建湿度限制因子来表达湿度对每一个树种生长的胁迫作用。

$$f_{\text{drought}} = \frac{4\text{DI} \cdot (\text{DI}_{\max} - \text{DI})}{\text{DI}_{\max}^2} \tag{7-12}$$

式中，f_{drought} 代表干旱对生长的胁迫作用；DI 代表研究区的干燥度指数；$\text{DI}_{\max}$ 代表树种 i 在其地理分布区范围内能够忍受的最大干燥度。

7.2.1.12 竞争限制因子

本研究构建一个密度制约竞争因子，即拥挤效应指数来代表邻体竞争对树木生长的胁迫作用(Botkin et al. 1972，葛剑平 1996)。

$$f_{\text{competition}} = 1 - \frac{W_{\text{tot}}}{W_{\max}} \tag{7-13}$$

式中，$f_{\text{competition}}$ 代表竞争限制因子；$W_{\max}$ 代表以邻体半径作圆的空间范围内所能承受的最大的生物量；W_{tot} 代表以邻体半径作圆的空间范围内邻体的总生物量。

$$W_{\text{tot}} = \sum W_i \cdot \text{PA}_i \cdot \text{PDW}_i \tag{7-14}$$

式中，W_i 代表圆内邻体 i 的生物量；PA_i 代表基株与邻体 i 之间欧氏空间距离加权系数；PDW_i 代表基株与邻体 i 之间系统发育距离加权系数。

第 3 章与第 5 章研究发现，在群落水平个体尺度下存在显著的负密度制约死亡与生长效应，说明典型阔叶红松林存在种内种间竞争差异(图 3-2、图 5-1、图 5-2)。以往研究通常混淆了种内竞争和种间竞争。本研究使用基株与邻体之间的种内或种间系统发育距离对总生物量指数进行加权修正，以表示考虑了种间竞争差异。

$$\text{PDW}_i = 1 - \left[(1 - \text{PDW}_0) \left(1 - \text{Exp}\left(\frac{-3 \cdot \text{PD}}{\text{PD}_0} \right) \right) \right] \tag{7-15}$$

式中，PDW_i 代表系统发育距离权重；PDW_0 代表最小权重值(0.6)；PD 代表基株与邻体之间系统发育距离；PD_0 代表超过此系统发育距离阈值(600)时，系统发育距离权重逐渐趋于平缓。

PA_i代表基株与邻体之间的欧氏空间距离加权函数(Busing 1991):

$$PA_i = \frac{THE_i - \sin(THE_i)}{\pi} \tag{7-16}$$

式中,THE_i代表基株与邻体间的夹角,单位为弧度。

$$THE_i = 2 \cdot \arccos\left(\frac{Dist_i}{R}\right) \tag{7-17}$$

式中,$Dist_i$代表基株与邻体之间的欧氏空间距离;R代表邻体影响半径。

第3章研究发现,邻体半径种间变异对密度制约效应研究结果造成影响,因此不可忽略树种邻体半径种间变异事实(图3-4),因此本研究对此进行修正。TMDS模型中邻体半径由基株树种与坡向(山脊、山坡、山谷)共同决定。以往大多数的树木死亡过程模型或是森林动态林隙模型中的死亡亚模型,通常对所有树种的树木无差别地采用统一的、固定的邻体影响半径(Smith and Urban 1988, Busing 1991, Canham et al. 2004)。但是,正如第3章研究结果所示,若不考虑邻体半径的种间和环境变异将对密度制约研究产生错误判断。不同树种应有不同的邻体半径,邻体半径的大小将影响树种对生物邻体效应的解释。

7.2.1.13 树木生长模块

在植被动态模型中,树木死亡一般被认为是树木生长效率和树木年龄的函数。近年来,有学者证实树木年龄应替换成树木的大小,一般由胸径来表示(Manusch et al. 2012)。Botkin 等(1972)在 JABOWA 模型中,假设树木未能维持某个最小年生长量时,树木的死亡率就会大幅升高。Leemans (1991)在其研制的 FORSKA 模型中,也借鉴了该思想,并提出了一个生长效率的概念,认为树木死亡与其生长效率息息相关。

TMDS 模型中的树木生长模块借鉴了 FORSKA 模型中的算法。该模块由两部分构成:树木不同垂直层次上叶片总光合同化量的累计;树干边材呼吸消耗量(邵国凡 1991, Bugmann 2001)。该模块以1年为时间步长,每年模拟计算每株林木的胸径生长、树高生长及叶面积生长等。

$$\frac{d(D^2H)}{dt} = f(\text{Environment})\int_B^H S_L(\gamma P_z - \delta z)dz \tag{7-18}$$

式中,D代表胸径;H代表树高;B代表枝下高;S_L代表叶面积密度;γ代表某特定树种的总光合速率,P_z代表某特定树种树梢向下 z m 处的光合作用响应值;δ代表某特定树种边材维持成本;$f(\text{Environment})$代表总的生长限制因子。

生长方程中有两个重要的参数需要说明：

(1) 光合作用常数(γ)：生物学意义为树木在无非光合组织的呼吸消耗条件下，单位时间单位树冠高度产生的叶面积数量，单位为 $cm^2/(m \cdot a)$。

(2) 呼吸作用常数(δ)：生物学意义为活的非光合组织为了维持养分和水分运输所消耗的能量，在生长方程中具体化为单位时间单位面积边材所导致的叶面积减少的数量，单位为 $cm^2/(m^2 \cdot a)$。

大多数基于过程的树木死亡模型通常采用假设相对简单且稍显粗糙的树冠结构，将每株林木的整个树冠视为平面的圆盘或者圆柱体，而此假设忽略了针叶树种和阔叶树种之间树冠结构形态上的差异(Botkin et al. 1972, Busing 1991, Leemans 1991, Yan and Shugart 2005)。TMDS 模型与其他林隙模型的主要区别为树冠结构的刻画更加细致。本研究将针叶树种的树冠结构近似刻画成圆锥体，而将阔叶树种的树冠结构近似刻画成椭球体(Nilson 1999)。因此，相应地修改 FORSKA 模型中的叶面积密度函数，由单位冠高叶面积修改为单位树冠体积叶面积。

$$S_L = \frac{L}{V} \tag{7-19}$$

式中，L 和 V 分别代表整个树冠的总叶面积和体积。

参考 FORSKA 模型，建立光合作用效应函数，P_z 是光强对树木生长影响的比例因子，取值范围 0~1，是无量纲因子。

$$P_z = \frac{kI_z - c}{\alpha - c} \tag{7-20}$$

式中，k 代表群落消光系数；I_z 代表树梢向下 z m 处的光强；α 代表光半饱和点(LSP)；c 代表光补偿点(LCP)。

树梢向下 z m 处的光强的减少主要受到上层树冠两个方面的影响：一方面是上层树冠叶量；另一方面是上层树冠的叶倾角分布(leaf inclination angle distribution, LAD)。上层树冠叶量对光强的消减作用由叶面积指数(leaf area index, LAI)表示，上层树冠的叶倾角对光强的消减作用采用群落消光系数表示。在单株林木树冠，光强从上而下逐渐减弱，其分布服从 Lambert-Beer 定律(Monsi and Saeki 1953)：

$$I_z = I_0 e^{-k \cdot LAI_z} \tag{7-21}$$

式中，I_z 代表树梢向下 z m 处的光强；I_0 代表树梢处光强；LAI_z 代表树梢向下 z m 处的叶面积指数；k 代表群落消光系数。

相似地，某基株树梢处的光强由基株所属样方上部的光强及树高高于基株的所有邻体总的距离加权叶面积共同决定，光强分布依然遵循 Lambert-Beer 定律（Monsi and Saeki 1953）：

$$I_0 = I_{quad} e^{-k \cdot LA_t} \tag{7-22}$$

式中，I_{quad} 代表基株所属样方上部的光强；LA_t 代表邻体半径作圆的范围内，树高高于基株的所有邻体总的欧氏距离加权叶面积。

$$LA_t = \sum LA_{ti} \cdot PA_i \tag{7-23}$$

式中，LA_{ti} 代表邻体半径作圆的范围内，高于基株的邻体 i 总叶面积；PA_i 代表基株与邻体 i 之间的欧氏距离加权系数。

第 4 章研究发现，生境异质性对树木死亡存在影响，生境异质性与树种生境偏好可共同导致生境过滤效应，促进物种共存，不可忽略生境异质性对样地空间资源的重新分配作用。然而，以往的植被动态模型通常将研究区假设为一个均质的物理环境（Bugmann 2001），因此本研究对此进行修正。本研究考虑了生境异质性对太阳辐射强度重新分配的作用，采用山体阴影来表征生境异质性对样地内生长季平均入射光强水平分布的影响。山体阴影是一个地形因子，能够表征某一区域（如样方）的相对光照强度（Burrough and McDonell 1998）。

$$I_{quad} = \frac{Hillshade_{quad}}{Hillshade_{mean}} \cdot I_{mean} \tag{7-24}$$

式中，I_{mean} 代表研究区生长季平均入射光强；$Hillshade_{quad}$ 代表某特定样方的山体阴影值；$Hillshade_{mean}$ 代表研究区山体阴影平均值。

7.2.1.14 外部干扰死亡

风倒木风险死亡率与树种是否具有浅根性及土壤湿度有关。

$$U_2 = WMP_0 \cdot \left[1 - \exp\left(\frac{-3 \cdot SM}{SM_0} \right) \right] \tag{7-25}$$

式中，U_2 代表外部干扰死亡率；WMP_0 代表每年最大风倒死亡率（0.03）；SM 代表土壤湿度；SM_0 代表超过此土壤湿度阈值（50%）时，树木外部干扰死亡率逐渐趋于平缓。

模型中的更多其他详情，如胸径生长、树高生长、叶面积生长等，以往研究已有大量介绍，请参见 FORSKA 模型查看相关细节。

7.2.2 模型检验与参数敏感性分析

模型检验包括确证性检验和有效性检验。模型分析包括模型结构分析、参数敏感性分析及模型不确定性分析等。

模型的检验是模型模拟研究中必不可少的组成部分，是模型模拟结果和适宜性的基础与保证。目前，树木死亡过程模型的精度验证由于以往研究往往缺乏树木死亡动态监测数据而导致难度较大(桑卫国和李景文 1998, 国庆喜和葛剑平 2007)。近年来，大型森林动态监测样地的出现和发展为树木死亡过程模型的有效性检验提供了有力的数据支持。

概括性、真实性和精确性是衡量生态模型成功与否的主要判别准则。经验模型，即统计模型，是根据系统变量之间回归关系建立的模型，它们通常具有精确性和真实性，但是概括性不强，模型的应用范围较窄，并且需要大量的树木生长数据。机理模型具有概括性和真实性，但缺乏精确性，它们比抽象模型需要更多的调查数据，但比经验模型需要用到的调查数据要少(葛剑平 1996)。

对于森林动态模型的检验一般从两个方面入手：其一是看它的子模型是否符合逻辑；其二是看整体模型在模拟中的表现是否符合事实规律(左金淼 2004)。目前，过程模型的检验方式主要包括确证性(verification)检验和有效性(validation)检验(Oreskes et al. 1994, Edward and Rykiel 1996, Loehle 1997, Purschke et al. 2013)。确证性检验主要关注如何将过程模型的结构特征正确地转化为程序代码及计算机程序调试等。Nathan 等(2001)将确证性检验定义为模型的总体表现，可包括模型模拟结果与实际观测在空间分布、趋势走向、变化过程等方面的吻合程度；有效性检验定义为利用精度评价指标定量地对模型模拟值与观测值进行比较，机理模型是一个理论的探索过程，不能像对经验模型那样苛求于模型对细节刻画的精准程度。本研究采用确证性检验(定性)与有效性检验(定量)相结合的方式对模型进行验证。

模型检验包括定性检验和定量检验，即确证性检验和有效性检验。定性检验是比较模拟的树木死亡空间格局与样地实测的树木死亡空间格局的吻合程度。定量检验是比较预测树木死亡率与观测树木死亡率之间的差异，使用模型误差进行量化评价。

$$\text{Error}(\%) = \frac{\text{PredMort} - \text{ObsMort}}{\text{ObsMort}} \cdot 100\% \tag{7-26}$$

式中，PredMort 为模型模拟 50 次的树木死亡率预测值，ObsMort 为树木死亡率观测值。

此外，本研究对树木生长也进行检验，因为树木生长受压死亡强烈依赖于冠层竞争。本研究采用平均误差（mean error, ME）来量化评价树木生长过程。

$$\mathrm{ME} = \sum_{i=1}^{N} \frac{\mathrm{PredGrowth}_i - \mathrm{ObsGrowth}_i}{N} \tag{7-27}$$

式中，$\mathrm{PredGrowth}_i$ 为第 i 株树木生长实际观测值；$\mathrm{ObsGrowth}_i$ 为第 i 株树木生长模型预测值；N 为树木株数。

参数敏感性分析：输出变量对输入参数的敏感性作为参数敏感性分析的度量指标（$\Delta Y/\Delta X$），即参数 X 对输出变量 Y 的效应。敏感性度量指标（$\Delta Y/\Delta X$）为输出变量对输入参数的变化率（%）（Tatarinov and Cienciala 2006）。敏感度分为三个等级：①高敏感参数：$|\Delta Y/\Delta X| \geq 0.5$；②中敏感参数：$0.3 \leq |\Delta Y/\Delta X| < 0.5$；③低敏感参数：$|\Delta Y/\Delta X| < 0.3$。候选待检验敏感参数包括：$PD_0$、$PDW_0$、$SurvYrs$、$WMP_0$ 及 SM_0。敏感性计算公式如下：

$$\mathrm{Sensitivity} = \frac{\Delta Y}{\Delta X} \tag{7-28}$$

式中，ΔY 为输出变量变化率；ΔX 输出参数变量率。设定 6 种输出参数变化率：0.05、0.10、0.20、–0.05、–0.10 及–0.20。

此外，计算比较在不同模型设置（即系统发育加权与不加权、固定邻体半径与变化邻体半径、空间异质与空间同质、圆柱体与圆锥体/椭球体树冠构筑型）下的模型误差，以此判定模型结构改进是否提高了模型模拟精度。

7.2.3 模拟程序设计与实现

JABOWA、FORSKA、SPACE 等经典森林动态模拟模型通常采用 Fortran 语言编写，本研究研制的树木死亡动态模拟 TMDS 模型是在 Visual Studio 2010（Microsoft, Redmond, WA, USA）集成开发环境下，采用 C#编程语言编写，并引用 GDAL/OGR 1.10.1（http://www.gdal.org/）地理信息系统开源库及 R.NET 1.5.13（https://rdotnet.codeplex.com/）开源库。GDAL/OGR 是一个在 X/MIT 许可协议下的开源栅格/矢量空间数据转换库，目前几乎所有 GIS 类产品（如 ArcGIS、Google Earth）内核均使用了 GDAL/OGR 来读写空间数据。模拟程序采用面向对象程序设计，应用多种设计模式（design patterns），以达到易维护、易扩展、易复用、灵活性好的目的（图 7-2）。每株树在样地内的地理信息及属性采用 ESRI Shapefile 矢量数据格式存储；每个样方的山体阴影及土壤湿度采用 GeoTIFF 栅格数据格式存储。程序主界面包括开始、数据预处理、参数估计、模拟、帮助等一级菜单，以及树冠投影面积计算、树木矢量数据生成、地形因子计算、地形因子赋值、树高曲线方程、叶面积经验方程、树干生物量经验方程、树木死亡动态模拟、模型参考文献、关于程序等二级菜单（图 7-3、图 7-4）。

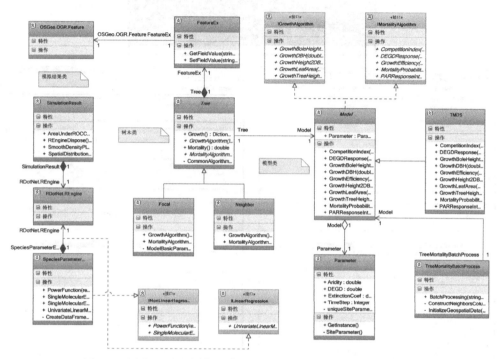

图 7-2 树木死亡动态模拟主程序主要统一建模语言(UML)类图
Fig. 7-2 Main unified modelling language (UML) class diagram of tree mortality dynamic simulator

除此之外，程序还可实现根据带有树木坐标及属性的 MS Excel 表格文件生成 ESRI Shapefile 点状矢量文件；根据附带有树木 8 个方向冠幅信息的 MS Excel 表格文件生成 ESRI Shapefile 树冠投影多边形矢量文件并计算其面积；根据 GeoTIFF 格式的 DEM 栅格文件生成坡度、坡向、凹凸度及山体阴影等 GeoTIFF 格式的地形栅格文件，并将各地形因子的栅格值赋予相应的树木；通过 C#与 R 混编方式，实现对树高、叶面积、树干生物量分别与胸径之间非线性回归关系建立统计模型，并自动调用 R 函数进行相应的参数估计；在程序中可直接查看 Adobe PDF 格式的模型主要参考文献。模型程序核心源代码请参见附录。

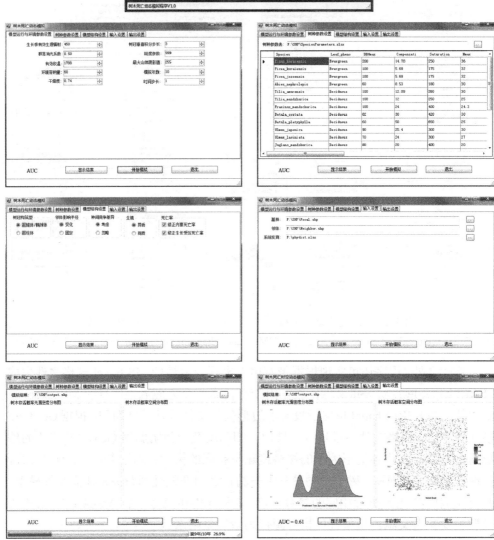

图 7-3 树木死亡动态模拟程序主界面、模拟和运行结果界面

Fig. 7-3 Main form, model simulation and running interface of tree mortality dynamic simulator

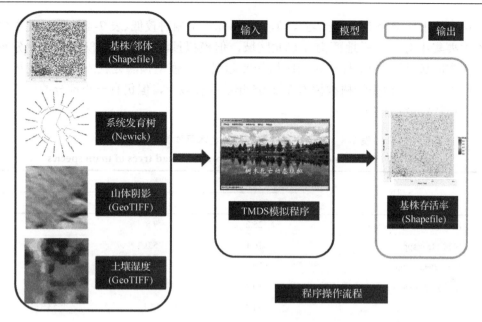

图 7-4 树木死亡动态模拟程序操作流程
Fig. 7-4 Procedures of tree mortality dynamic simulator

7.3 结　　果

7.3.1 确证性检验

先锋树种和伴生树种（桦、杨、槭）的实测死亡率相对较高（表 7-4）。虽然红松

表 7-4　主要树种实际观测和模型预测死亡率对比
Table 7-4　Observed vs. predicted tree mortality rates of main species

树种	观测死亡率/%	预测死亡率（平均）[*]/%	误差/%[**]
冷杉 *Abies nephrolepis*	6.8	6.8	0.0
槭树 *Acer* spp.	6.9	5.2	−24.6
桦树 *Betula* spp.	8.9	14.0	+57.3
水曲柳 *Fraxinus mandshurica*	1.9	3.2	+68.4
云杉 *Picea* spp.	11.4	10.2	−10.5
红松 *Pinus koraiensis*	3.5	2.8	−20.0
杨树 *Populus* spp.	19.6	23.2	+18.4
椴树 *Tilia* spp.	3.7	5.2	+40.5
榆树 *Ulmus* spp.	4.1	5.6	+36.6
所有树种	5.7	5.1	−10.5

*　模型共模拟运行 50 次。
**　正值代表模型高估，负值代表模型低估。

死亡木的平均胸径最大(表 7-5),但其实际死亡率相对较低(表 7-4)。实际死亡树木主要集中分布在样地西南和东北区域,但模拟预测死亡树木仅主要集中分布在西南区域(图 7-5)。样地东南区域的实际观测和模型预测死亡木均分布较稀疏(图 7-5)。大多数先锋树种树木在 2005 年的胸径较小,但仍有一小部分的径级较大、年龄较老(图 7-6)。

表 7-5 主要树种实际观测死亡木平均胸径与范围
Table 7-5 Mean and range of DBH for observed dead trees of main species

树种	胸径/cm		
	平均值	最小值	最大值
冷杉 Abies nephrolepis	25.2	9.8	45.0
槭树 Acer spp.	16.1	5.0	30.5
桦树 Betula spp.	20.9	4.5	69.0
水曲柳 Fraxinus mandschurica	30.3	6.7	47.8
云杉 Picea spp.	34.2	17.1	53.3
红松 Pinus koraiensis	48.3	12.0	77.0
杨树 Populus spp.	22.1	4.5	133.0
椴树 Tilia spp.	24.8	10.2	53.0
榆树 Ulmus spp.	23.7	7.8	49.3
所有树种	27.2	4.2	133.0

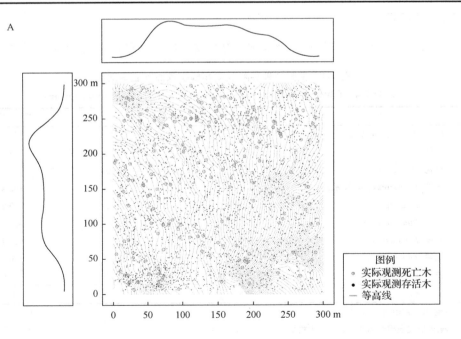

7 物种共存理论修正树木死亡过程模型 · 87 ·

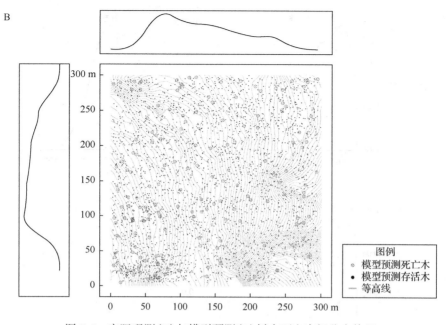

图 7-5 实际观测(A)与模型预测(B)树木死亡空间分布格局
Fig. 7-5 Spatial distribution pattern of observed (A) vs. predicted (B) tree mortality

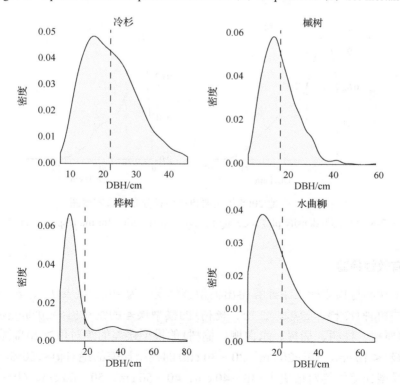

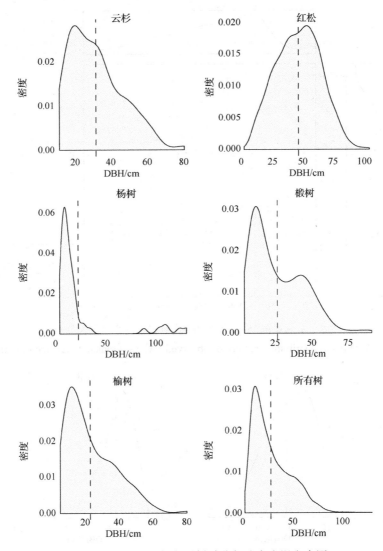

图 7-6 样地 2005 年主要树种胸径分布光滑密度图

Fig. 7-6 Smooth density plot showing the DBH distribution for main species in 2005

7.3.2 有效性检验

基于观测与预测死亡率计算得出模型整体误差为−10.5%(表 7-4、表 7-6),其中,某些树种(冷杉、云杉、红松、杨树)的模型误差的绝对值≤20%,但是,其他树种(槭树、桦树、椴树、水曲柳、榆树)的模型误差的绝对值＞20%(表 7-4)。较小径级(＜10 cm、10～20 cm、20～30 cm)的模型误差的绝对值＜20%,而其他径级的模型误差的绝对值更大(30～40 cm、40～50 cm、50～60 cm、60～70 cm、

≥70 cm)(表 7-6)。

表 7-6 不同径级实际观测和模型预测死亡率对比
Table 7-6 Observed vs. predicted tree mortality rates for different DBH classes

径级/cm	观测死亡率/%	预测死亡率(平均)*/%	误差**/%
<10	10.1	8.8	−12.9
10~20	5.6	6.6	+17.9
20~30	4.8	4.2	−12.5
30~40	5.3	3.8	−28.3
40~50	4.8	2.5	−47.9
50~60	4.2	2.7	−35.7
60~70	5.3	4.1	−22.6
≥70	5.9	9.3	+57.6
所有径级	5.7	5.1	−10.5

* 模型共模拟运行 50 次。
** 正值代表模型高估，负值代表模型低估。

对于模型树木生长部分的评价，所有树木整体的平均误差为−0.19 cm/a(表 7-7、表 7-8)。其中，某些树种((槭树、桦树、红松、椴树)的平均误差的绝对值<0.20 cm/a，而其他树种的平均误差的绝对值更大(表 7-7)。某些径级(<10 cm、10~20 cm、60~70 cm、≥70 cm)的平均误差的绝对值<0.20 cm/a，而其他径级的平均误差的绝对值更大(表 7-8)。

表 7-7 主要树种的树木生长评价
Table 7-7 Tree growth evaluation for main species

树种	平均误差*/(cm/a)
冷杉 *Abies nephrolepis*	−0.24
槭树 *Acer* spp.	−0.13
桦树 *Betula* spp.	−0.13
水曲柳 *Fraxinus mandschurica*	−0.22
云杉 *Picea* spp.	−0.17
红松 *Pinus koraiensis*	−0.48
杨树 *Populus* spp.	−0.15
椴树 *Tilia* spp.	−0.22
榆树 *Ulmus* spp.	−0.40
所有树种	−0.19

* 模型共模拟运行 50 次。正值代表模型高估，负值代表模型低估。

表 7-8 不同径级的树木生长评价
Table 7-8 Tree growth evaluation for different DBH classes

径级/cm	平均误差*/(cm/a)
<10	−0.15
10~20	−0.19
20~30	−0.22
30~40	−0.23
40~50	−0.22
50~60	−0.20
60~70	−0.17
≥70	−0.02
所有径级	−0.19

*模型共模拟运行 50 次。正值代表模型高估，负值代表模型低估。

7.3.3 参数敏感性分析

参数敏感性分析显示，参数 PD_0 和 SM_0 有 2(2/6)个高敏感等级，参数 WMP_0 有 3(3/6)个高敏感等级，参数 PDW_0 有 4(4/6)个高敏感等级，而参数 SurvYrs 有 6(6/6)个高敏感等级(表 7-9)。模型采用空间同质性和圆柱树冠构筑型时的模型误差(−7%)的绝对值小于模型缺省设置(−10.5%)(表 7-10)。模型采用无系统发育加权和固定邻体半径时的模型误差(−12.2%、+19.3%)的绝对值大于模型缺省设置(−10.5%)(表 7-10)。当模型忽略任何一种树木死亡模块(内禀、生长受压和风倒木死亡)时，都将造成树木死亡率的低估(表 7-10)。

表 7-9 参数敏感性分析
Table 7-9 Parameter sensitivity analysis

参数	缺省参数值	参数变化率	输出变化率(平均)*	敏感性	敏感等级
PD_0	0.6	0.05	0.0784	1.5686	高
	0.6	0.10	−0.1765	−1.7647	高
	0.6	0.20	−0.0196	−0.0980	低
	0.6	−0.05	0.0196	−0.3922	中
	0.6	−0.10	0.0392	−0.3922	中
	0.6	−0.20	0.0392	−0.1961	低
PDW_0	600	0.05	0.0980	1.9608	高
	600	0.10	0.0588	0.5882	高
	600	0.20	0.0000	0.0000	低
	600	−0.05	0.1373	−2.7451	高
	600	−0.10	0.0588	−0.5882	高
	600	−0.20	−0.0784	0.3922	中

续表

参数	缺省参数值	参数变化率	输出变化率(平均)*	敏感性	敏感等级
WMP_0	0.03	0.05	0.0392	0.7843	高
	0.03	0.10	0.0588	0.5882	高
	0.03	0.20	0.0000	0.0000	低
	0.03	−0.05	−0.0784	1.5686	高
	0.03	−0.10	−0.0196	0.1961	低
	0.03	−0.20	0.0392	−0.1961	低
SM_0	50	0.05	−0.0392	−0.7843	高
	50	0.10	0.0196	0.1961	低
	50	0.20	−0.0784	−0.3922	中
	50	−0.05	0.0980	−1.9608	高
	50	−0.10	−0.0392	0.3922	中
	50	−0.20	−0.0784	0.3922	中
SurvYrs	20** 60*** 40****	0.05	−0.0980	−1.9608	高
	20** 60*** 40****	0.10	−0.3137	−3.1373	高
	20** 60*** 40****	0.20	−0.2745	−1.3725	高
	20** 60*** 40****	−0.05	−0.0392	0.7843	高
	20** 60*** 40****	−0.10	0.0784	−0.7843	高
	20** 60*** 40****	−0.20	0.3725	−1.8627	高

* 模型共模拟运行 50 次。** 先锋树种。*** 演替末期树种。**** 其他树种。

表 7-10　不同参数设置条件下的实际观测与模拟预测死亡率
Table 7-10　Observed vs. predicted tree mortality rates for different settings

模型设置	观测死亡率/%	预测死亡率(平均)*/%	误差**/%
缺省设置	5.7	5.1	−10.5
无系统发育加权	5.7	5.0	−12.2
固定邻体半径	5.7	6.8	+19.3
空间同质性	5.7	5.3	−7.0
圆柱树冠	5.7	5.3	−7.0
无风倒木死亡率	5.7	4.4	−22.8
无内禀死亡率	5.7	2.9	−49.1
无生长受压死亡率	5.7	1.3	−77.2

* 模型共模拟运行 50 次。缺省设置是指模型同时采用系统发育加权、变化的邻体半径、空间异质性、圆锥体/椭球体树冠构筑型、内禀死亡、生长受压死亡、风倒木死亡。
** 正值代表模型高估，负值代表模型低估。

7.4 讨 论

7.4.1 树木死亡空间格局

从实际观测与模型预测死亡木空间格局的结果来看，死亡木在样地西南区域的分布较为集中(图 7-5)，这可能是由于该区域曾经受过轻度干扰所致(徐丽娜和金光泽 2012)。枫桦、白桦和山杨属于阔叶红松林中的先锋树种，同时也是喜光树种，在所有被研究的树种中具有较高的光补偿点。光补偿点是净光合速率为零时的光强，其值越低说明叶片利用弱光的能力越强。先锋树种通常首先侵入并定植于受干扰地区，通过降低其林下演替末期树种生长率或存活率的方式来抑制其更新(He and Duncan 2000)。另外，由于同种邻体或者亲缘关系较近的邻体之间又具有相同或相似的资源需求，从而导致强烈的种内或种间竞争(Burns and Strauss 2011)，致使树木的生长效率降低，树木死亡率升高，即同种负密度制约死亡和系统发育负密度制约死亡。生物邻体间相互作用发生的种内、种间竞争，尤其是同种或近缘物种之间对光资源的竞争，导致某些树木光合能力下降，因碳饥饿而死亡。通常耐阴树种的树冠较厚，而喜光树种的冠长较薄、冠幅较窄，导致喜光树种具有强烈的自我稀疏(self-thinning)。再者，先锋树种通常生长较快而寿命相对较短，具有较高的内禀死亡率(Manusch et al. 2012)。实际上，先锋树种(桦、杨)实际观测树木死亡率也相对较高(表 7-4)。大多数先锋树种树木在 2005 年的胸径较小，但仍有一小部分的径级较大、年龄较老(图 7-6)。

从实际观测死亡木空间格局的结果来看，死亡木在样地东北区域的分布也较为集中，但是与模型预测结果并不完全吻合(图 7-5)。模型对所有树木进行预测的误差为-10.5%(表 7-4)。青楷槭和花楷槭均属于阔叶红松林中的伴生树种，其光补偿点虽大于红松，但小于先锋树种。这两个槭属树种主要聚集分布在样地东北部地势平坦地区的林下层(徐丽娜和金光泽 2012)，而在林冠层与亚冠层均几乎没有分布，光照强度受到林冠层和亚冠层树木的遮挡而导致林下层光照条件相对较差(李志宏 2009)。树木之间对光照的竞争依赖于树木及其树冠的空间分布。最近有学者在美国温带森林的研究结果发现，光照是影响树木生长最为重要的因子(McMahon et al. 2011)。与先锋树种枫桦、白桦与山杨相似，亲缘关系较近的树种(如槭属树种青楷槭和花楷槭)由于具有相同或相似的资源需求，导致其种内、种间竞争较为强烈。树木邻体对基株生长、存活的负效应主要表现为两个方面：一个方面是光资源的直接竞争；另一方面是高大邻体树冠对低矮基株的遮挡和消光作用，如 Lambert-Beer 定律所描述和强调的(Monsi and Saeki 1953)。再者，当树木越接近甚至到达特定树种的最大胸径时，即达到老龄树时，树木的内禀死亡率也会随之升高。而由于青楷槭和花楷槭这两个槭属树种的最大胸径均相对较小(大约 30 cm)，因此，其内禀死亡率也会相对较高。

从实际观测与模型预测死亡木空间格局的结果来看，死亡木在样地东南区域的分布较为发散(图 7-5)，这可能是由于该地区集中分布着大量大径级红松。成年红松个体已经经历了生活史早期阶段的生境过滤作用并成功存活在适宜生境，并且占据了阔叶红松林中的林冠层，虽然红松幼年耐阴，而到了成年后较为喜光，但红松的光补偿点相对较低(Liu et al. 2014)。因此，成年红松之间对于光资源的竞争相对较弱，不易引发强烈的负密度制约死亡(外部胁迫死亡)。然而内禀死亡却可能是成年红松死亡的主要诱发因素。本样地以往研究认为红松生理上的衰老是其粗木质残体形成最重要的驱动因子(刘妍妍和金光泽 2010)。但是，由于红松是阔叶红松林中的建群种，并且其寿命较长(通常达 200～400 年)，最大胸径较大(最大可达 200 cm)，因此，在模型模拟的 10 年内树木死亡概率相对较低。根据 2015 年样地复查时所调查的 2005～2015 年间野外实测数据，虽然红松死亡木的平均胸径在所有研究树种中最大，但是其死亡木比例低于以上讨论的先锋树种和伴生树种(图 7-5)。

样地内共有两处地势较为陡峭的地区，分别位于东部和东南地区。具有大径级、长寿命、死亡率相对较低等特点的成年红松适宜生长在这些陡坡生境，而许多具有短寿命、喜光特征的先锋树种树木则多集中分布在样地的西南地区，并且往往具有相对较高的死亡率。此外，样地东部和东南地区的海拔均高于样地的西南区域(图 7-6)。但是，这并不能推出海拔是树木死亡的直接影响因素，因为海拔属于地形因子，是一种间接生态因子，海拔只能通过与它相关性较高的其他因素(如土壤属性)协变，间接地对树木死亡造成影响，正如 4.4.1 节所讨论的，样地内某些变异幅度较大的土壤属性可能随海拔而发生协变。红松虽然具有浅根性，但是侧根比较发达，适宜生长在山坡或山脊等土壤排水良好地区(如陡坡地区)。以往研究发现，成年红松常常聚集分布在样地的陡坡地区(徐丽娜和金光泽 2012)。生境异质性和特定树种的生境偏好能够引发树种空间分布呈现聚集格局状态(Piao et al. 2013)。经历生境过滤后存活下来的成年大树则生长在较适宜的生境(Russo et al. 2005)。本样地以往研究发现，由于较大尺度生境异质性的影响，样地内绝大多数主要树种均表现出了一定程度的生境偏好(Piao et al. 2013)。这与生活史早期阶段发生的生境过滤导致成年后树种和生境出现明显关联性的观点一致(Comita and Engelbrecht 2009)。由于模型模拟的数据集中，红松的平均胸径均较大(DBH＞30 cm 的比例高达 80%以上)，说明大多数红松在生活史早期阶段已经经历了生境过滤，并生长在环境适宜的陡坡地区。

7.4.2 模型结构、参数与输入的不确定性

树木死亡个体模型确证性检验显示，模型整体误差为-10.5%，但存在种间差

异和径级间差异(表7-4、表7-6)。对于树木生长模块评价结果显示,所有树木的平均误差为–0.19 cm/a (表7-7、表7-8),可见该树木死亡个体模型仍有较大的改进空间以增强模型表现。在森林群落动态中,树木死亡是最具有不确定性的部分,生长受压和由干扰引起的树木死亡通常在树木死亡个体中具有一定的不确定性(Bugmann et al. 2019)。该树木死亡个体模型的不确定性很可能来自模型结构、参数与输入三个部分。

(1) 模型结构的不确定性。森林动态模型中的生长受压死亡公式可分为三类:理论型、经验型、高度机理型(Bugmann et al. 2019)。本研究中的树木死亡个体的生长受压死亡公式衍生自FORSKA林窗模型,而林窗模型中的生长受压死亡公式一般基于理论假设(Bugmann et al. 2019)。理论型树木死亡公式对所作假设高度敏感(Bugmann 2001)。事先判定森林模型中哪一种理论型树木死亡公式正确,几乎不可能(Bugmann et al. 2019)。经验型树木死亡公式通常基于树轮宽度数据(以及基于轮宽的树木死亡预警信号指数)或森林清查数据(Mamet et al. 2015, Cailleret et al. 2019, Bugmann et al. 2019)。但是在森林动态长期模拟中,树木死亡经验公式的缺点是不同的死亡经验公式将导致巨大的森林群落演替差异(Bircher et al. 2015)。通过基于几十年的时间序列数据来鉴别一个合适的树木死亡公式通常是不可能的(Bugmann et al. 2019)。本研究所建立的树木死亡个体模型存在一些模型结构的不确定性。在不同模型设置的模型误差比较中可见,某些符合预期假设(如在模型竞争指数中引入系统发育加权和变化的邻体半径可提高模型模拟精度)(表7-10),但是也有些与预期不符(如空间异质性、复杂树冠构筑型的引入并未提高模型模拟精度)(表7-10)。

(2) 参数的不确定性。由于生态系统结构和功能的复杂性,以及当前对其了解的有限,致使参数不确定性几乎是不可避免的(Song et al. 2013)。本研究的树木死亡个体模型中存在一些参数假设。例如,在生长受压死亡公式中,基于生长率和寿命之间的种间权衡关系(Bigler and Veblen 2009, Rötheli et al. 2011),假设先锋树种、演替末期树种和其他树种的未校正的年生长受压死亡分别为 0.12、0.04、0.06(参数 SurvProb)。通过参数敏感性分析可知,参数 SurvYrs 可能是该分析所有参数中最敏感的参数(表7-9)。

(3) 输入数据的不确定性。树木死亡个体模型中的树木生长检验结果显示,树木生长模拟结果并不理想,平均误差为–0.19 cm/a(表7-7、表7-8)。而树木生长受压死亡强烈依赖于冠层光资源的竞争。因此,模型输入的不确定性很可能与冠层相关联的树木变量有关(如树高、枝下高、冠幅、叶面积等)。但是,野外调查时测量仪器(Vertex III测高器)的测量误差是不可避免的。

7.4.3 森林动态监测大样地对树木死亡个体模型发展的支持

大型森林动态监测样地可为林分尺度森林动态模拟及树木死亡个体模型提供模型初始化、参数化及模型验证的数据支持。树木死亡个体模型的发展(如模型结构、初始化、参数化及模型评价等)需要更多的数据支持，以及更好地理解树木死亡过程来改进模拟未来森林动态的稳健性(Bugmann et al. 2019)。树木死亡个体模型未来的发展可能更加依赖于方式集成，并融合多种数据源(Bugmann et al. 2019)。

大型森林动态监测样地可作为一个重要的数据源支持林分尺度树木死亡个体模型的发展。①模型初始化：大型森林动态监测样地提供了树木初始信息，如树种、坐标、胸径、树高等，而以往研究由于缺少数据支持不得不多从裸地开始模拟(Shugart et al. 2018)。②模型参数化：大型森林动态监测样地的调查工作包括核心调查和补充调查两部分，核心调查包括树木生长、死亡、更新动态、树木坐标、地形因子等，补充调查包括植物功能性状、土壤、气候数据的测量等，上述数据是生态系统模型以及地球系统模型进行参数化非常宝贵的数据源(Anderson-Teixeira et al. 2015)，而上述模型通常均为基于个体(如 ED/ED2 模型)(Moorcroft et al. 2001; Medvigy et al. 2009)。③模型评价：以往对于树木死亡模型的检验多在全球或区域尺度上进行，利用大型森林动态监测样地数据在林分尺度上验证此类模型的研究鲜有报道。近年来研究显示，林窗模型已被应用于森林群落构建研究当中(Chauvet et al. 2017)，但应用于大型森林动态监测样地的研究甚少。

7.4.4 展望

该树木死亡个体模型仍存在一定局限性。①模型检验：虽然本研究由 9 hm^2 的森林动态监测样地 2005~2015 年 10 年间树木死亡动态监测提供数据支撑，样地面积远大于大多数以往森林动态模拟模型研究所采用的样地面积大小(通常为 1 hm^2)(葛剑平 1996, 桑卫国和李景文 1998)，但树木死亡的监测时间跨度可能仍有待进一步增加，因为树木死亡是一个长期、复杂的生态过程。国内外发展较早且相对比较成熟的大型森林动态监测样地，例如，巴拿马 BCI 热带雨林动态大样地从 1980 年建成并首次调查以来(Hubbell and Foster 1983, Condit 1998)，共计复查达 7 次之多，已具有近 40 年的长期动态监测历史。随着森林动态监测样地的进一步发展，它不仅能为森林群落物种共存与生物多样性维持机制研究提供良好的科研平台，同时也必将为森林动态模拟模型提供长期、有效的模型验证数据。本研究是森林动态模拟模型中的树木死亡子模型集成森林动态监测样地及物种共存机制的一次尝试，以期为以后的研究提供一种新的思路和角度。②树冠结构参数：由于传统测量方法的限制，目前模型所模拟的树木均为树高>10 m 的树木，尚未包含低矮小树,近年来出现并迅速发展的轻型无人机(unmanned aerial vehicle, UAV)可为局域尺度监测提供有力支持。无人机为长期生态监测提供廉价而高分辨的原始数据(Zhang et al. 2016a)。未来研究可以使用无人机或三脚架为平台、激光雷达(Light detection and ranging, LiDAR)作为传感器,精确测量样地内树冠的水平和垂直结构参数,如胸径、树高、枝下高、树冠形态参数等。该技术的使用有望克服传统测量方法的弱点,对低矮小树的树冠形态以及树高等参数进行精确测量。树冠形态参数对基于过程的机理模型中准确计算叶片遮蔽作用及光合作用至关重要(Bugmann 2001)。③生理参数:近年来,侧重树木生理生态过程,细致刻画了碳饥饿和水力失衡假说的树木死亡过程模型(如 CASTANEA 模型)有新的发展(Davi and Cailleret 2017);Farquhar 碳同化生物化学模型对光合作用也有着细致的刻画,值得借鉴。④因为过程模型和经验模型均存在不足之处,因此以后的模型研制过程中,可以考虑构建"混合"模型的方式(hybrid modelling approach),即半机理半统计模型,将两种建模方式的优点充分发挥,已有若干树木生长和死亡模型尝试此种建模方式(Valentine and Mäkelä 2005, Waterworth et al. 2007, Girardin et al. 2008, Adams et al. 2013)。C#与 R 混合编程可为实现半机理半统计模型提供技术保障。⑤模型中仅采用了气候平均值,未考虑到年际变异。

7.5 本章小结

根据前文关于典型阔叶红松林物种共存理论的研究结果,对经典树木死亡过程模型的部分结构进行了修正,并在 Visual Studio 集成开发环境下采用 C#语言编程实现,集成森林动态监测样地数据,以 1 年为时间步长,对研究区 2005～2015 年间树木死亡进行模拟并检验。结果表明,模型整体误差为–10.5%,且误差存在种间差异和径级间差异。本研究是森林动态模拟研究集成森林动态监测样地及物种共存理论的一次尝试,以期为以后的相关研究提供一种新思路和新角度。

参 考 文 献

蔡慧颖. 2012. 森林火灾损失评估方法的研究. 哈尔滨: 东北林业大学硕士学位论文.

曹坤芳, 常杰. 2010. 突发气象灾害的生态效应: 2008年中国南方特大冰雪灾害对森林生态系统的破坏. 植物生态学报, 34(2): 123-124.

陈志成, 万贤崇. 2016. 虫害叶损失造成的树木非结构性碳减少与树木生长-死亡的关系研究进展. 植物生态学报, 40(9): 958-968.

储诚进, 王酉石, 刘宇, 等. 2017. 物种共存理论研究进展. 生物多样性, 25(4): 345-354.

董蕾, 李吉跃. 2013. 植物干旱胁迫下水分代谢、碳饥饿与死亡机理. 生态学报, 33(18): 5477-5483.

杜春英, 李帅, 刘丹, 等. 2010. 大兴安岭地区森林雷击火发生的时空分布. 自然灾害学报, 19(3): 72-76.

葛剑平. 1996. 森林生态学建模与仿真. 哈尔滨: 东北林业大学出版社.

葛结林, 熊高明, 邓龙强, 等. 2012. 湖北神农架山地米心水青冈-多脉青冈混交林的群落动态. 生物多样性, 20(6): 643-653.

国庆喜, 葛剑平. 2007. 基于个体的集水区森林动态模型. 应用生态学报, 18(8): 1778-1784.

韩大校, 金光泽. 2017. 地形和竞争对典型阔叶红松林不同生长阶段树木胸径生长的影响. 北京林业大学学报, 39(1): 9-19.

金毅, 陈建华, 米湘成, 等. 2015. 古田山24 ha森林动态监测样地常绿阔叶林群落结构和组成动态: 探讨2008年冰雪灾害的影响. 生物多样性, 23(5): 610-618.

李景文. 1997. 红松混交林生态与经营. 哈尔滨: 东北林业大学出版社.

李俊清, 牛树奎, 刘艳红. 2017. 森林生态学(第三版). 北京: 高等教育出版社.

李志宏. 2009. 阔叶红松林主要组成树种树冠特征及其对更新的影响. 哈尔滨: 东北林业大学硕士学位论文.

梁玉莲. 2010. 干扰对东北次生林结构与过程影响的模拟体系研究. 哈尔滨: 东北林业大学硕士学位论文.

刘妍妍, 金光泽. 2010. 小兴安岭阔叶红松林粗木质残体基础特征. 林业科学, 46(4): 8-14.

任思远, 王婷, 祝燕, 等. 2014. 暖温带-北亚热带过渡带落叶阔叶林群落不同径级系统发育结构的变化. 生物多样性, 22(5): 574-582.

桑卫国, 陈灵芝, 马克平. 1999. 蒙古栎红松林演替模型FOROAK的研究. 植物学报, 41(6): 658-668.

桑卫国, 李景文. 1998. 小兴安岭南坡红松林动态模拟. 生态学报, 18(1): 38-47.

邵国凡. 1991. 阔叶红松林主要树种生长的水热指数和最优生长过程的模拟研究. 林业科学, 27(1): 21-27.

汪殷华, 米湘成, 陈声文, 等. 2011. 古田山常绿阔叶林主要树种2002—2007年间更新动态. 生物多样性, 19(2): 178-189.

王慧杰, 常顺利, 张毓涛, 等. 2016. 天山雪岭云杉森林群落的密度制约效应. 生物多样性, 24(3): 252-261.

王业蘧. 1995. 阔叶红松林. 哈尔滨: 东北林业大学出版社.

徐丽娜, 金光泽. 2012. 小兴安岭凉水典型阔叶红松林动态监测样地: 物种组成与群落结构. 生物多样性, 20(4): 470-481.

杨光. 2006. 东北东部山区部分主要树种邻体影响半径研究. 哈尔滨: 东北林业大学硕士学位论文.

张觅, 米湘成, 金光泽. 2014. 小兴安岭凉水谷地云冷杉林群落组成与空间格局. 科学通报, 59(24): 2377-2387.

张泽浦, 方精云, 菅诚. 2000. 邻体竞争对植物个体生长速率和死亡概率的影响: 基于日本落叶松种群试验的研究. 植物生态学报, 24(3): 340-345.

赵雪, 徐丽娜, 金光泽. 2015. 地形对典型阔叶红松林灌木更新的影响. 生物多样性, 23(6): 767-774.

左金淼. 2004. 阔叶红松林动态模拟的研究. 北京: 北京林业大学硕士学位论文.

Ackerly D D, Cornwell W K. 2007. A trait-based approach to community assembly: Partitioning of species trait values into within- and among-community components. Ecology Letters, 10(2): 135-145.

Adams H D, Williams A P, Xu C, et al. 2013. Empirical and process-based approaches to climate-induced forest mortality models. Frontiers in Plant Science, 4: 1-5.

Adler J. 2012. R in a Nutsbell, 2nd Edition. Sebastopol O'Reilly Media.

Adler P B, Fajardo A, Kleinhesselink A R, et al. 2013. Trait-based tests of coexistence mechanisms. Ecology Letters, 16(10): 1294-1306.

Anderegg W R, Callaway E S. 2012. Infestation and hydraulic consequences of induced carbon starvation. Plant Physiology, 159(4): 1866-1874.

Anderegg W R, Hicke J A, Fisher R A, et al. 2015. Tree mortality from drought, insects, and their interactions in a changing climate. New Phytologist, 208(3): 674-683.

Anderson-Teixeira K J, Davies S J, Bennett A C, et al. 2015. CTFS-ForestGEO: a worldwide network monitoring forests in an era of global change. Global Change Biology, 21(2): 528-549.

Bai X, Queenborough S A, Wang X, et al. 2012. Effects of local biotic neighbors and habitat heterogeneity on tree and shrub seedling survival in an old-growth temperate forest. Oecologia, 170(3): 755-765.

Baldeck C A, Harms K E, Yavitt J B, et al. 2017. Habitat filtering across tree life stages in tropical forest communities. Proceedings of the Royal Society B: Biological Sciences, 280(1766): 20130548.

Barigah T S, Charrier O, Douris M, et al. 2013. Water stress-induced xylem hydraulic failure is a causal factor of tree mortality in beech and poplar. Annals of Botany, 112(7): 1431-1437.

Basille M, Calenge C, Marboutin É, et al. 2008. Assessing habitat selection using multivariate statistics: some refinements of the ecological-niche factor analysis. Ecological Modelling, 211(1-2): 233-240.

Bates D, Maechler M, Bolker B, et al. 2015. Fitting linear mixed-effects models using lme4. Journal of Statistical Software, 67: 1-48.

Biging G S, Dobbertin M. 1992. A comparison of distance-dependent competition measures for height and basal area growth of individual conifer trees. Forest Science, 38(3): 695-720.

Bigler C, Bugmann H. 2004. Predicting the time of tree death using dendrochronological data. Ecological Applications, 14(3): 902-914.

Bigler C, Veblen T T. 2009. Increased early growth rates decrease longevities of conifers in subalpine forests. Oikos, 118(8): 1130-1138.

Bircher N, Cailleret M, Bugmann H. 2015. The agony of choice: different empirical mortality models lead to sharply different future forest dynamics. Ecological Applications, 25(5): 1303-1318.

Bivand R, Yu D. 2015. spgwr: Geographically weighted regression. R package version 0.6-28.

Boardman N K. 1977. Comparative photosynthesis of sun and shade plants. Annual Review of Plant Physiology, 28: 355-377.

Bolker B M, Brooks M E, Clark C J, et al. 2009. Generalized linear mixed models: a practical guide for ecology and evolution. Trends in Ecology and Evolution, 24(3): 127-135.

Bolmgren K, Cowan P D. 2008. Time-size tradeoffs: a phylogenetic comparative study of flowering time, plant height and seed mass in a north-temperate flora. Oikos, 117(5): 424-429.

Bossel H. 1986. Dynamics of forest dieback: systems analysis and simulation. Ecological Modelling, 34(3-4): 259-288.

Botkin D B, Janak J F, Wallis J R. 1972. Some ecological consequences of a computer model of forest growth. Journal of Ecology, 60(3): 849-872.

Boyden S, Binkley D, Shepperd W. 2005. Spatial and temporal patterns in structure, regeneration, and mortality of an old-growth ponderosa pine forest in the Colorado Front Range. Forest Ecology and Management, 219(1): 43-55.

Bremer B, Bremer K, Chase M, et al. 2009. An update of the Angiosperm Phylogeny Group classification for the orders and families of flowering plants: APG III. Botanical Journal of the Linnean Society, 161(2): 105-121.

Brodribb T J, Cochard H. 2009. Hydraulic failure defines the recovery and point of death in water-stressed conifers. Plant Physiology, 149(1): 575-584.

Bugmann H K M, Yan X, Sykes M T, et al. 1996. A comparison of forest gap models: model structure and behaviour. Climatic Change, 34(2): 289-313.

Bugmann H, Seidl R, Hartig F, et al. 2019. Tree mortality submodels drive simulated long-term forest dynamics: assessing 15 models from the stand to global scale. Ecosphere, 10(2): e02616.

Bugmann H. 2001. A review of forest gap models. Climatic Change, 51(3-4): 259-305.

Burnham K P, Anderson D R. 2002. Model selection and multimodel inference: a practical information-theoretic approach. New York: Springer Science & Business Media.

Burns J H, Strauss S Y. 2011. More closely related species are more ecologically similar in an experimental test. Proceedings of the National Academy of Sciences of the United States of America, 108(13): 5302-5307.

Burrough P A, McDonell R A. 1998. Principles of geographical information systems. New York: Oxford University Press.

Busing R T. 1991. A spatial model of forest dynamics. Vegetatio, 92(2): 167-179.

Cailleret M, Dakos V, Jansen S, et al. 2019. Early-warning signals of individual tree mortality based on annual radial growth. Frontiers in Plant Science, 9(1964): 1-14.

Cailleret M, Jansen S, Robert E M, et al. 2017. A synthesis of radial growth patterns preceding tree mortality. Global Change Biology, 23(4): 1675-1690.

Canham C D, LePage P T, Coates K D. 2004. A neighborhood analysis of canopy tree competition: effects of shading versus crowding. Canadian Journal of Forest Research, 34(4): 778-787.

Canham C D, Papaik M J, Latty E F. 2001. Interspecific variation in susceptibility to windthrow as a function of tree size and storm severity for northern temperate tree species. Canadian Journal of Forest Research, 31(1): 1-10.

Chauvet M, Kunstler G, Roy J, et al. 2017. Using a forest dynamics model to link community assembly processes and traits structure. Functional Ecology, 31(7): 1452-1461.

Chen J, Bradshaw G A. 1999. Forest structure in space: a case study of an old growth spruce-fir forest in Changbaishan Natural Reserve, PR China. Forest Ecology and Management, 120(1-3): 219-233.

Chen L, Comita L S, Wright S J, et al. 2017. Forest tree neighborhoods are structured more by negative conspecific density dependence than by interactions among closely related species. Ecography, 41(7): 1114-1123.

Chen Y, Wright S J, Muller-Landau H C, et al. 2016. Positive effects of neighborhood complementarity on tree growth in a Neotropical forest. Ecology, 97(3): 776-785.

Chesson P. 2000. Mechanisms of maintenance of species diversity. Annual Review of Ecology and Systematics, 31: 343-366.

Chi X, Tang Z, Xie Z, et al. 2015. Effects of size, neighbors, and site condition on tree growth in a subtropical evergreen and deciduous broad-leaved mixed forest, China. Ecology and Evolution, 5(22): 5149-5161.

Choat B. 2013. Predicting thresholds of drought-induced mortality in woody plant species. Tree Physiology, 33(7): 669-671.

Coley P D. 1988. Effects of plant growth rate and leaf lifetime on the amount and type of anti-herbivore defense. Oecologia, 74(4): 531-536.

Comita L S, Condit R, Hubbell S P. 2007. Developmental changes in habitat associations of tropical trees. Journal of Ecology, 95(3): 482-492.

Comita L S, Engelbrecht B M. 2009. Seasonal and spatial variation in water availability drive habitat associations in a tropical forest. Ecology, 90(10): 2755-2765.

Comita L S, Hubbell S P. 2009. Local neighborhood and species' shade tolerance influence survival in a diverse seedling bank. Ecology, 90(2): 328-334.

Comita L S, Muller-Landau H C, Aguilar S, et al. 2010. Asymmetric density dependence shapes species abundances in a tropical tree community. Science, 329(5989): 330-332.

Comita L S, Queenborough S A, Murphy S J, et al. 2014. Testing predictions of the Janzen-Connell hypothesis: a meta-analysis of experimental evidence for distance- and density-dependent seed and seedling survival. Journal of Ecology, 102(4): 845-856.

Condit R, Ashton P S, Baker P, et al. 2010. Spatial patterns in the distribution of tropical tree species. Science, 288(5470): 1114-1119.

Condit R, Hubbell S P, Foster R B. 1992. Recruitment near conspecific adults and the maintenance of tree and shrub diversity in a Neotropical Forest. The American Naturalist, 140(2): 261-286.

Condit R. 1998. Tropical forest census plots. Berlin: Springer-Verlag.

Connell J H. 1971. On the role of natural enemies in preventing competitive exclusion in some marine animals and in rain forest trees. Netherlands: Centre for Agricultural Publication and Documentation, Wageningen.

Coomes D A, Allen R B. 2007. Mortality and tree-size distributions in natural mixed-age forests. Journal of Ecology, 95(1): 27-40.

Csilléry K, Kunstler G, Courbaud B, et al. 2017. Coupled effects of wind-storms and drought on tree mortality across 115 forest stands from the Western Alps and the Jura mountains. Global Change Biology, 23(12): 5092-5107.

Darwin A T, Ladd D, Galdins R, et al. 2004. Response of forest understory vegetation to a major ice storm. Journal of the Torrey Botanical Society, 131: 45-52.

Das S, De M, Ray R, et al. 2012. Microbial ecosystem in Sunderban Mangrove Forest Sediment, north-east coast of Bay of Bengal, India. Geomicrobiology Journal, 29(7): 656-666.

Davi H, Cailleret M. 2017. Assessing drought-driven mortality trees with physiological process-based models. Agricultural and Forest Meteorology, 232: 279-290.

De'ath G. 2002. Multivariate regression trees: a new technique for modeling species-environmental relationships. Ecology, 83(4): 1105-1117.

Díaz S, Kattge J, Cornelissen J H, et al. 2016. The global spectrum of plant form and function. Nature, 529(7585): 167-171.

Diggle P J. 2003. Statistical analysis of point processes. London: Academic Press.

Druckenbrod D L, Shugart H H, Davies I. 2005. Spatial pattern and process in forest stands within the Virginia piedmont. Journal of Vegetation Science, 16(1): 37-48.

Du Y, Queenborough S A, Chen L, et al. 2017. Intraspecific and phylogenetic density-dependent seedling recruitment in a subtropical evergreen forest. Oecologia, 184(1): 193-203.

Edward J, Rykiel J. 1996. Testing ecological models: the meaning of validation. Ecological Modelling, 90(3): 229-244.

Fan C, Tan L, Zhang P, et al. 2017. Determinants of mortality in a mixed broad-leaved Korean pine forest in northeastern China. European Journal of Forest Research, 136(3): 457-469.

Fielding A H, Bell J F. 1997. A review of methods for the assessment of prediction errors in conservation presence/absence models. Environmental Conservation, 24: 38-49.

Foster A C, Shuman J K, Shugart H H, et al. 2017. Validation and application of a forest gap model to the southern Rocky Mountains. Ecological Modelling, 351: 109-128.

Fotheringham A S, Brunsdon C, Charlton M. 2003. Geographically weighted regression: the analysis of spatially varying relationships. New York: John Wiley & Sons.

Franklin J F, Shugart H, Harmon M E. 1987. Tree death as an ecological process. BioScience, 37(8): 550-556.

Galle A, Esper J, Feller U, et al. 2010. Responses of wood anatomy and carbon isotope composition of *Quercus pubescens* saplings subjected to two consecutive years of summer drought. Annals of Forest Science, 67(8): 809-809.

Gause G F. 1934. The struggle for existence. Baltimore: Williams & Wilkins.

Gaylord M L, Kolb T E, Pockman W T, et al. 2013. Drought predisposes pinon-juniper woodlands to insect attacks and mortality. The New Phytologist, 198(2): 567-578.

Ge J, Li J, Guo H, et al. 1995. Study on community structure and self-sustaining mechanism of broad-leaved Korean pine forests in northeast China. Journal of Northeast Forestry University, 6(3): 11-18.

Ge J. 1994. Study of structure and dynamics of Korean pine forest in the Xiaoxing'an Mountains. Journal of Northeast Forestry University, 5(4): 1-5.

Gelman A, Hill J. 2006. Data analysis using regression and multilevel/hierarchical models. Cambridge: Cambridge University Press.

Girardin M P, Raulier F, Bernier P Y, et al. 2008. Response of tree growth to a changing climate in boreal central Canada: a comparison of empirical, process-based, and hybrid modelling approaches. Ecological Modelling, 213(2): 209-228.

Gollini I, Lu B, Charlton M, et al. 2015. GWmodel: an R package for exploring spatial heterogeneity using geographically weighted models. Journal of Statistical Software, 63(117): 1-50.

Gravel D, Canham C D, Beaudet M, et al. 2010. Shade tolerance, canopy gaps and mechanisms of coexistence of forest trees. Oikos, 119(3): 475-484.

Gray L, He F. 2009. Spatial point-pattern analysis for detecting density-dependent competition in a boreal chronosequence of Alberta. Forest Ecology and Management, 259(1): 98-106.

Grayson L M, Progar R A, Hood S M. 2017. Predicting post-fire tree mortality for 14 conifers in the Pacific Northwest, USA: Model evaluation, development, and thresholds. Forest Ecology and Management, 399: 213-226.

Greenwood S, Ruiz-Benito P, Martinez-Vilalta J, et al. 2017. Tree mortality across biomes is promoted by drought intensity, lower wood density and higher specific leaf area. Ecology Letters, 20(4): 539-553.

Grimm V, Berger U, Bastiansen F, et al. 2006. A standard protocol for describing individual-based and agent-based models. Ecological Modelling, 198(1-2): 115-126.

Grimm V, Berger U, DeAngelis D L, et al. 2010. The ODD protocol: A review and first update. Ecological Modelling, 221(23): 2760-2768.

Grubb P J. 1977. The maintenance of species-richness in plant communities: the importance of the regeneration niche. Biological Review, 52(1): 107-145.

Guo L, Ma Z, Zhang L. 2008. Comparison of bandwidth selection in application of geographically weighted regression: a case study. Canadian Journal of Forest Research, 38(9): 2526-2534.

Hamilton D A. 1986. A logistic model of mortality in thinned and unthinned mixed conifer stands of northern Idaho. Forest Science, 32(4): 989-1000.

Hamilton G J. 1969. The dependence of volume increment of individual trees on dominance, crown dimensions, and competition. Forestry, 42(2): 133-144.

Harms K E, Condit R, Hubbell S P, et al. 2001. Habitat associations of trees and shrubs in a 50‐ha Neotropical forest plot. Journal of Ecology, 89(6): 947-959.

Harms K E, Wright S J, Calderón O, et al. 2000. Pervasive density-dependent recruitment enhances seedling diversity in a tropical forest. Nature, 404(6777): 493-495.

Hartmann H. 2011. Will a 385 million year-struggle for light become a struggle for water and for carbon? - How trees may cope with more frequent climate change-type drought events. Global Change Biology, 17(1): 642-655.

He F, Duncan R P. 2000. Density‐dependent effects on tree survival in an old-growth Douglas fir forest. Journal of Ecology, 88(4): 676-688.

HilleRisLambers J, Adler P, Harpole W, et al. 2012. Rethinking community assembly through the lens of coexistence theory. Annual Review of Ecology, Evolution, and Systematics, 43: 227-248.

Hirzel A H, Hausser J, Chessel D, et al. 2002. Ecological-niche factor analysis: how to compute habitat-suitability maps without absence data? Ecology, 83(7): 2027-2036.

Hosmer D W, Lemeshow S. 2000. Applied logistic regression. New York: John Wiley & Sons, Inc.

Hubbell S P, Foster R B. 1983. Diversity of canopy trees in a Neotropical forest and implications for conservation. Tropical rain forest: ecology and management. Blackwell Scientific, Oxford.

Hubbell S P, Foster R B. 1992. Short-term dynamics of a neotropical forest: why ecological research matters to tropical conservation and management. Oikos, 63(1): 48-61.

Hubbell S P. 1979. Tree dispersion, abundance, and diversity in a tropical dry forest. Science, 30(4387): 1299-1309.

Hülsmann L, Bugmann H, Brang P. 2017. How to predict tree death from inventory data-lessons from a systematic assessment of European tree mortality models. Canadian Journal of Forest Research, 47(7): 890-900.

Hutchinson G E. 1957. Concluding remarks. Cold Spring Harbor Symposia on Quantitative Biology, 22: 415-427.

Iida Y, Kohyama T S, Swenson N G, et al. 2014a. Linking functional traits and demographic rates in a subtropical tree community: the importance of size dependency. Journal of Ecology, 102(3): 641-650.

Iida Y, Poorter L, Sterck F, et al. 2014b. Linking size-dependent growth and mortality with architectural traits across 145 co-occurring tropical tree species. Ecology, 95(2): 353-363.

Jacquet J S, Bosc A, O'Grady A, et al. 2014. Combined effects of defoliation and water stress on pine growth and non-structural carbohydrates. Tree Physiology, 34(4): 367-376.

Janzen D H. 1970. Herbivores and the number of tree species in tropical forests. The American Naturalist, 104(940): 501-528.

Johnson D J, Beaulieu W T, Bever J D, et al. 2012. Conspecific negative density dependence and forest diversity. Science, 336(6083): 904-907.

Johnson D J, Bourg N A, Howe R, et al. 2014. Conspecific negative density-dependent mortality and the structure of temperate forests. Ecology, 95(9): 2493-2503.

King D A, Davies S J, Noor N S M. 2006. Growth and mortality are related to adult tree size in a Malaysian mixed dipterocarp forest. Forest Ecology and Management, 223(1-3): 152-158.

Kitajima K, Poorter L. 2008. Functional basis for resource niche partitioning by tropical trees. Tropical Forest Community Ecology. Oxford: Blackwell Publishing.

Kitajima K. 1994. Relative importance of photosynthetic traits and allocation patterns as correlates of seedling shade tolerance of 13 tropical trees. Oecologia, 98(3-4): 419-428.

Korzukhin M D, Ter-Mikaelian M T, Wagner R G. 1996. Process versus empirical models: which approach for forest ecosystem management? Canadian Journal of Forest Research, 26(5): 879-887.

Kosola K R, Dickmann D I, Paul E A, et al. 2001. Repeated insect defoliation effects on growth, nitrogen acquisition, carbohydrates, and root demography of poplars. Oecologia, 129(1): 65-74.

Kushwaha C P, Tripathi S K, Singh K P. 2010. Tree specific traits affect flowering time in Indian dry tropical forest. Plant Ecology, 212(6): 985-998.

Lai J, Mi X, Ren H, et al. 2009. Species-habitat associations change in a subtropical forest of China. Journal of Vegetation Science, 20(3): 415-423.

Lan G, Getzin S, Wiegand T, et al. 2012. Spatial distribution and interspecific associations of tree species in a tropical seasonal rain forest of China. PLoS One, 7(9): e46074.

Lan G, Hu Y, Cao M, et al. 2011. Topography related spatial distribution of dominant tree species in a tropical seasonal rain forest in China. Forest Ecology and Management, 262(8): 1507-1513.

Lavoie M, Harper K, Paré D, et al. 2007. Spatial pattern in the organic layer and tree growth: A case study from regenerating *Picea mariana* stands prone to paludification. Journal of Vegetation Science, 18(2): 213-222.

Lebrija-Trejos E, Wright SJ, Hernández A, et al. 2014. Does relatedness matter? Phylogenetic density-dependent survival of seedlings in a tropical forest. Ecology, 95(4): 940-951.

Leemans R, Prentice I C. 1987. Description and simulation of tree-layer composition and size distributions in a primaeval *Picea-Pinus* forest. Vegetatio, 69(1): 147-156.

Leemans R. 1991. Sensitivity analysis of a forest succession model. Ecological Modelling, 53: 247-262.

Lemon P C. 1961. Forest ecology of ice storms. Bulletin of the Torrey Botanical Club, 88(1): 21-29.

Levins R. 1966. The strategy of model building in population ecology. American Scientist, 54(4): 421-431.

Li H, Hoch G, Körner C. 2002. Source/sink removal affects mobile carbohydrates in *Pinus cembra* at the Swiss treeline. Trees, 16: 331-337.

Lin F, Comita L S, Wang X, et al. 2014. The contribution of understory light availability and biotic neighborhood to seedling survival in secondary versus old-growth temperate forest. Plant Ecology, 215(8): 795-807.

Lin L, Comita L S, Zheng Z, et al. 2012. Seasonal differentiation in density-dependent seedling survival in a tropical rain forest. Journal of Ecology, 100(4): 905-914.

Lin Y C, Comita L S, Johnson D J, et al. 2017. Biotic vs abiotic drivers of seedling persistence in a tropical karst forest. Journal of Vegetation Science, 28(1): 206-217.

Liu Q, Bi L, Song G, et al. 2018. Species-habitat associations in an old-growth temperate forest in northeastern China. BMC Ecology, 18(1): 20.

Liu Y, Li F, Jin G. 2014. Spatial patterns and associations of four species in an old-growth temperate forest. Journal of Plant Interactions, 9(1): 745-753.

Loehle C. 1997. A hypothesis testing framework for evaluating ecosystem model performance. Ecological Modelling, 97(3): 153-165.

Lorimer C G. 1983. Tests of age-independent competition indices for individual trees in natural hardwood stands. Forest Ecology and Management, 6(4): 343-360.

Lu J, Johnson D J, Qiao X, et al. 2015. Density dependence and habitat preference shape seedling survival in a subtropical forest in central China. Journal of Plant Ecology, 8(6): 568-577.

Lugo A E, Scatena F N. 1996. Background and catastrophic tree mortality in tropical moist, wet, and rain forests. Biotropica, 28(4): 585-599.

Lutz J A, Halpern C B. 2006. Tree mortality during early forest development: a long-term study of rates, causes, and consequences. Ecological Monographs, 76(2): 257-275.

MacArthur R H, Levins R. 1964. Competition, habitat selection and character displacement. Proceedings of the National Academy of Sciences of the United States of America, 51(6): 1207-1210.

Maguire D A, Hann D W. 1989. The relationship between gross crown dimensions and sapwood area at crown base in Douglas-fir. Canadian Journal of Forest Research, 19(5): 557-565.

Mamet S D, Chun K P, Metsaranta J M, et al. 2015. Tree rings provide early warning signals of jack pine mortality across a moisture gradient in the southern boreal forest. Environmental Research Letters, 10(8): 084021.

Mantgem P J V, Stephenson N L, Byrne J C, et al. 2009. Widespread increase of tree mortality rates in the western United States. Science, 323(5913): 521-524.

Manusch C, Bugmann H, Heiri C, et al. 2012. Tree mortality in dynamic vegetation models-A key feature for accurately simulating forest properties. Ecological Modelling, 243: 101-111.

McDowell N G. 2011. Mechanisms linking drought, hydraulics, carbon metabolism, and vegetation mortality. Plant Physiology, 155(3): 1051-1059.

McDowell N, Pockman W T, Allen C D, et al. 2008. Mechanisms of plant survival and mortality during drought: why do some plants survive while others succumb to drought? New Phytologist, 178(4): 719-739.

McMahon S M, Metcalf C J E, Woodall C W. 2011. High-dimensional coexistence of temperate tree species: functional traits, demographic rates, life history stages, and their physical context. PLoS One, 6(1): e16253.

Medvigy D, Wofsy S C, Munger J W, et al. 2009. Mechanistic scaling of ecosystem function and dynamics in space and time: ecosystem demography model version 2. Journal of Geophysical Research, 114(G1): G01002.

Mencuccini M, Martinez-Vilalta J, Hamid H A, et al. 2007. Evidence for age- and size-mediated controls of tree growth from grafting studies. Tree Physiology, 27(3): 463-473.

Mencuccini M, Martínez-Vilalta J, Vanderklein D, et al. 2005. Size-mediated ageing reduces vigour in trees. Ecology Letters, 8(11): 1183-1190.

Mencuccini M, Oñate M, Peñuelas J, et al. 2014. No signs of meristem senescence in old Scots pine. Journal of Ecology, 102(3): 555-565.

Messaoud Y, Houle G. 2006. Spatial patterns of tree seedling establishment and their relationship to environmental variables in a cold-temperate deciduous forest of eastern North America. Plant Ecology, 185(2): 319-331.

Metz M R. 2012. Does habitat specialization by seedlings contribute to the high diversity of a lowland rain forest? Journal of Ecology, 100(4): 969-979.

Monserud R A, Ledermann T, Sterba H. 2004. Are self-thinning constraints needed in a tree-specific mortality model? Forest Science, 50(6): 848-858.

Monsi M, Saeki T. 1953. Über den Lichtfaktor in den Pflanzengesellschaften und seine Bedeutung für die Stoffproduktion (On the factor light in plant communities and its importance for matter production). Japanese Journal of Botany, 14: 22-52.

Moorcroft P R, Hurtt G C, Pacala S W. 2001. A method for scaling vegetation dynamics: the ecosystem demography model (ED). Ecological Monographs, 71(4): 557-586.

Moore I D, Grayson R B, Landson A R. 1991. Digital terrain modelling: a review of hydrological, geomorphological, and biological applications. Hydrological Processes, 5(1): 3-30.

Mueller-Dombois D. 1987. Natural dieback in forests. BioScience, 37(8): 575-583.

Munné-Bosch S. 2008. Do perennials really senesce? Trends in Plant Science, 13(5): 216-220.

Munné-Bosch S. 2015. Senescence: is it universal or not? Trends in Plant Science, 20(11): 713-720.

Nakaya T, Fotheringham A, Brunsdon C, et al. 2005. Geographically weighted Poisson regression for disease association mapping. Statistics in Medicine, 24(17): 2695-2717.

Nakaya T, Fotheringham A, Charlton M, et al. 2009. Semiparametric geographically weighted generalised linear modelling in GWR 4.0.

Nathan R, Safriel U N, Noy-Meir I. 2001. Field validation and sensitivity analysis of a mechanistic model for tree seed dispersal by wind. Ecology, 82: 374-388.

Nilson T. 1999. Inversion of gap frequency data in forest stands. Agricultural and Forest Meteorology, 98: 437-448.

Nilson T. Kuusk A. 2004. Improved algorithm for estimating canopy indices from gap fraction data in forest canopies. Agricultural and Forest Meteorology, 124(3-4): 157-169.

Oreskes N, Shrader-Frechette K, Belitz K. 1994. Verification, validation, and confirmation of numerical models in the earth sciences. Science, 263(f147): 641-646.

Pacala S W, Canham C D, Saponara J, et al. 1996. Forest models defined by field measurements: estimation, error analysis and dynamics. Ecological Monographs, 66(1): 1-43.

Paine C E, Norden N, Chave J, et al. 2012. Phylogenetic density dependence and environmental filtering predict seedling mortality in a tropical forest. Ecology Letters, 15(1): 34-41.

Parker J, Patton R L. 1975. Effects of drought and defoliation on some metabolites in roots of black oak seedlings. Canadian Journal of Forest Research, 5(3): 457-463.

Pedersen B S. 1998. The role of stress in the mortality of midwestern oaks as indicated by growth prior to death. Ecology, 79(1): 79-93.

Peng C, Ma Z, Lei X, et al. 2011. A drought-induced pervasive increase in tree mortality across Canada's boreal forests. Nature Climate Change, 1(9): 467-471.

Penuelas J, Munné-Bosch S. 2010. Potentially immortal? New Phytologist, 187(3): 564-567.

Pérez-Harguindeguy N, Díaz S, Garnier E, et al. 2013. New handbook for standardised measurement of plant functional traits worldwide. Australian Journal of Botany, 61(3): 167-234.

Phillips O L, Aragão L E O C, Lewis S L, et al. 2009. Drought sensitivity of the Amazon rainforest. Science, 323(5919): 1344-1347.

Piao T, Comita L S, Jin G, et al. 2013. Density dependence across multiple life stages in a temperate old-growth forest of northeast China. Oecologia, 172(1): 207-217.

Poorter L, Wright S J, Paz H, et al. 2008. Are functional traits good predictors of demographic rates? Evidence from five Neotropical forests. Ecology, 89(7): 1908-1920.

Pu X, Zhu Y, Jin G. 2017. Effects of local biotic neighbors and habitat heterogeneity on seedling survival in a spruce-fir valley forest, northeastern China. Ecology and Evolution, 7(13): 4582-4591.

Purschke O, Schmid B C, Sykes M T, et al. 2013. Contrasting changes in taxonomic, phylogenetic and functional diversity during a long-term succession: insights into assembly processes. Journal of Ecology, 101: 857-866.

Rees M, Condit R, Crawley M, et al. 2001. Long-term studies of vegetation dynamics. Science, 293(5530): 650-655.

Riginos C, Milton S J, Wiegand T. 2005. Context-dependent interactions between adult shrubs and seedlings in a semi-arid shrubland. Journal of Vegetation Science, 16(3): 331-340.

Ripley B D. 1977. Modeling spatial pattern. Journal of the Royal Statistical Society, Series B, 39(2): 172-212.

Rose K E, Atkinson R L, Turnbull L A, et al. 2009. The costs and benefits of fast living. Ecology Letters, 12(12): 1379-1384.

Rötheli E, Heiri C, Bigler C. 2011. Effects of growth rates, tree morphology and site conditions on longevity of Norway spruce in the northern Swiss Alps. European Journal of Forest Research, 131(4): 1117-1125.

Rüger N, Huth A, Hubbell S P, et al. 2009. Response of recruitment to light availability across a tropical lowland rain forest community. Journal of Ecology, 97(6): 1360-1368.

Rüger N, Huth A, Hubbell S P, et al. 2011. Determinants of mortality across a tropical lowland rainforest community. Oikos, 120(7): 1047-1056.

Russo S E, Davies S J, King D A, et al. 2005. Soil-related performance variation and distributions of tree species in a Bornean rain forest. Journal of Ecology, 93(5): 879-889.

Russo S E, Jenkins K L, Wiser S K, et al. 2010. Interspecific relationships among growth, mortality and xylem traits of woody species from New Zealand. Functional Ecology, 24(2): 253-262.

Ryan M G, Binkley D, Fownes J H. 1997. Age-related decline in forest productivity: pattern and process. Advances in Ecological Research, 27: 213-262.

Ryan M G. 2011. Tree responses to drought. Tree Physiology, 31(3): 237-239.

Sala A, Piper F, Hoch G. 2010. Physiological mechanisms of drought-induced tree mortality are far from being resolved. New Phytologist, 186(2): 274-281.

Schurr F M, Bossdorf O, Milton S J, et al. 2004. Spatial pattern formation in semi-arid shrubland: a priori predicted versus observed pattern characteristics. Plant Ecology, 173(2): 271-282.

Sevanto S, McDowell N G, Dickman L T, et al. 2014. How do trees die? A test of the hydraulic failure and carbon starvation hypotheses. Plant Cell and Environment, 37(1): 153-161.

Sharpe P J A. 1990. Forest modeling approaches: compromises between generality and precision. Portland: Timber Press.

Shugart H H, Wang B, Fischer R, et al. 2018. Gap models and their individual-based relatives in the assessment of the consequences of global change. Environmental Research Letters, 13(3): 033001.

Sillett S C, Pelt R V, Carrol A L, et al. 2015. How do tree structure and old age affect growth potential of California redwoods? Ecological Monographs, 85(2): 181-212.

Sillett S C, Pelt R V, Koch G W, et al. 2010. Increasing wood production through old age in tall trees. Forest Ecology and Management, 259(5): 976-994.

Silvertown J. 2004. Plant coexistence and the niche. Trends in Ecology and Evolution, 19(11): 605-611.

Smith T M, Urban D L. 1988. Scale and resolution of forest structural pattern. Vegetatio, 74(2): 143-150.

Song, X., Bryan, B.A., Almeida, A.C., et al. 2013. Time-dependent sensitivity of a process-based ecological model. Ecological Modelling 265: 114-123.

Stephenson N L, Mantgem P J V, Bunn A G, et al. 2011. Causes and implications of the correlation between forest productivity and tree mortality rates. Ecological Monographs, 81(4): 527-555.

Swenson N G, Erickson D L, Mi X, et al. 2012. Phylogenetic and functional alpha and beta diversity in temperate and tropical tree communities. Ecology, 93(8): S112-S125.

Tatarinov F A, Cienciala E. 2006. Application of BIOME-BGC model to managed forests: 1. Sensitivity analysis. Forest Ecology and Management, 237(1-3): 267-279.

Team R D C. 2015. R: a language and environment for statistical computing. Vienna: R Foundation for Statistical Computing.

Thomas S C. 1996. Relative size at onset of maturity in rain forest trees: a comparative analysis of 37 Malaysian species. Oikos, 76(1): 145-154.

Thornthwaite C W, Mather J R. 1957. Instructions and tables for computing potential evaporation and the water balance. Publications in Climatology, 10: 185-311.

Tilman D. 2004. Niche tradeoffs, neutrality, and community structure: a stochastic theory of resource competition, invasion, and community assembly. Proceedings of the National Academy of Sciences of the United States of America, 101(30): 10854-10861.

Tyree M T, Dixon M A. 1986. Water stress induced cavitation and embolism in some woody plants. Physiologia Plantarum, 66(3): 397-450.

Uriarte M, Canham C D, Thompson J, et al. 2004. A neighborhood analysis of tree growth and survival in a hurricane-driven tropical forest. Ecological Monographs, 74(4): 591-614.

Uriarte M, Clark J S, Zimmerman J K, et al. 2012. Multidimensional trade-offs in species responses to disturbance: implications for diversity in a subtropical forest. Ecology, 93(1): 191-205.

Uriarte M, Swenson N G, Chazdon R L, et al. 2010. Trait similarity, shared ancestry and the structure of neighbourhood interactions in a subtropical wet forest: implications for community assembly. Ecology Letters, 13(12): 1503-1514.

Valencia R, Foster R B, Villa G, et al. 2004. Tree species distributions and local habitat variation in the Amazon: large forest plot in eastern Ecuador. Journal of Ecology, 92(2): 214-229.

Valentine H, Mäkelä A. 2005. Bridging process-based and empirical approaches to modeling tree growth. Tree Physiology, 25(7): 769-779.

Vandermeer J H. 1972. Niche theory. Annual Review of Ecology and Systematics, 3(1): 107-132.

Venables W N, Ripley B D. 2002. Modern applied statistics with S. 4th Edition. New York: Springer.

Villar R, Robleto J R, Jong Y D, et al. 2006. Differences in construction costs and chemical composition between deciduous and evergreen woody species are small as compared to differences among families. Plant, Cell and Environment, 29(8): 1629-1643.

Visser M D, Bruijning M, Wright S J, et al. 2016. Functional traits as predictors of vital rates across the life cycle of tropical trees. Functional Ecology, 30(2): 168-180.

Vowels K M. 2012. Ice storm damage to upland oak-hickory forest at Bernheim Forest, Kentucky. The Journal of the Torrey Botanical Society, 139(4): 406-415.

Vuong Q H. 1989. Likelihood ratio tests for model selection and non-nested hypotheses. Econometrica, 57(2): 307-333.

Wang X, Comita L S, Hao Z, et al. 2012. Local-scale drivers of tree survival in a temperate forest. PLoS One, 7(2): e29469.

Waterworth R M, Richards G P, Brack C L, et al. 2007. A generalised hybrid process-empirical model for predicting plantation forest growth. Forest Ecology and Management, 238(1): 231-243.

Webb C O, Ackerly D D, McPeek M A, et al. 2002. Phylogenies and community ecology. Annual Review of Ecology and Systematics, 8(33): 475-505.

Webb C O, Donoghue M J. 2005. Phylomatic: tree assembly for applied phylogenetics. Molecular Ecology Notes, 5(1): 181-183.

Webb C O, Gilbert G S, Donoghue M J. 2006. Phylodiversity-dependent seedling mortality, size structure, and disease in a Bornean rain forest. Ecology, 87(s7): S123-S131.

Webster R, Oliver M A. 2007. Geostatistics for environmental scientists. New York: John Wiley & Sons.

Wheeler D. 2013. gwrr: Fits geographically weighted regression models with diagnostic tools. R package version 0.2-1.

Wiegand K, Jeltsch F, Ward D. 2000. Do spatial effects play a role in the spatial distribution of desert - dwelling Acacia raddiana? Journal of Vegetation Science, 11(4): 473-484.

Wiens J A. 1989. Spatial scaling in ecology. Functional Ecology, 3: 385-397.

Woolley T, Shaw D C, Ganio L M, et al. 2012. A review of logistic regression models used to predict post-fire tree mortality of western North American conifers. International Journal of Wildland Fire, 21(1): 1-5.

Wright S J, Carrasco C, Calderon O, et al. 1999. The Niño southern oscillation, variable fruit production, and famine in a tropical forest. Ecology, 80(5): 1632-1647.

Wright S J, Kitajima K, Kraft N J B, et al. 2010. Functional traits and the growth-mortality trade-off in tropical trees. Ecology, 91(12): 3664-3674.

Wright S J, Muller-Landau H C, Condit R, et al. 2003. Gap-dependent recruitment, realized vital rates, and size distributions of tropical trees. Ecology, 84(12): 3174-3185.

Wu H, Franklin S B, Liu J, et al. 2017. Relative importance of density dependence and topography on tree mortality in a subtropical mountain forest. Forest Ecology and Management, 384: 169-179.

Wu J, Swenson N G, Brown C, et al. 2016. How does habitat filtering affect the detection of conspecific and phylogenetic density dependence? Ecology, 97(5): 1182-1193.

Wu J. 2004. Effects of changing scale on landscape pattern analysis: scaling relations. Landscape Ecology, 19(2): 125-138.

Yan X, Shugart H H. 2005. FAREAST: a forest gap model to simulate dynamics and patterns of eastern Eurasian forests. Journal of Biogeography, 32(9): 1641-1658.

Zanne A E, Tank D C, Cornwell W K, et al. 2014. Three keys to the radiation of angiosperms into freezing environments. Nature, 506(7522): 89-92.

Zeileis A, Kleiber C, Jackman S. 2008. Regression models for count data in R. Journal of Statistical Software, 27(8): 1-15.

Zevenbergen L W, Thorne C R. 1987. Quantitative analysis of land surface topography. Earth Surface Processes and Landforms, 12(1): 47-56.

Zhang D Y, Zhang B Y, Lin K, et al. 2012a. Demographic trade-offs determine species abundance and diversity. Journal of Plant Ecology, 5(1): 82-88.

Zhang J, Hu J, Lian J, et al. 2016a. Seeing the forest from drones: testing the potential of lightweight drones as a tool for long-term forest monitoring. Biological Conservation, 198: 60-69.

Zhang J, Song B, Li B H, et al. 2010. Spatial patterns and associations of six congeneric species in an old-growth temperate forest. Acta Oecologia, 36(1): 29-38.

Zhang L W, Mi X C, Shao H B, et al. 2011. Strong plant-soil associations in a heterogeneous subtropical broad-leaved forest. Plant and Soil, 347(1): 211-220.

Zhang L, Gove J H. 2005. Spatial assessment of model errors from four regression techniques. Forest Science, 51(4): 334-346.

Zhang L, Ma Z, Guo L. 2009. An evaluation of spatial autocorrelation and heterogeneity in the residuals of six regression models. Forest Science, 55(6): 533-548.

Zhang X, Lei Y, Cai D, et al. 2012b. Predicting tree recruitment with negative binomial mixture models. Forest Ecology and Management, 270: 209-215.

Zhang Z, Papaik M J, Wang X, et al. 2016b. The effect of tree size, neighborhood competition and environment on tree growth in an old-growth temperate forest. Journal of Plant Ecology, 10(6): 970-980.

Zhang, X., Cao, Q.V., Duan, A., et al. 2017. Modeling tree mortality in relation to climate, initial planting density, and competition in Chinese fir plantations using a Bayesian logistic multilevel method. Canadian Journal of Forest Research 47(9): 1278-1285.

Zhen Z, Li F, Liu Z, et al. 2013. Geographically local modeling of occurrence, count, and volume of downwood in Northeast China. Applied Geography, 37(1): 114-126.

Zhu K, Woodall C, Monteiro J, et al. 2015a. Prevalence and strength of density-dependent tree recruitment. Ecology, 96(9): 2319-2327.

Zhu Y, Comita L S, Hubbell S P, et al. 2015b. Conspecific and phylogenetic density-dependent survival differs across life stages in a tropical forest. Journal of Ecology, 103(4): 957-966.

附录 树木死亡动态模拟(TMDS)模型程序核心源代码(C#)

```csharp
using System;
using System.Collections.Generic;
using System.Linq;
using System.Text;
using OSGeo.OGR;
using OSGeo.GDAL;
using OSGeo.OSR;
using System.Windows.Forms;
using System.Collections;
using System.Data;
using System.IO;
using RDotNet;
using System.Drawing;


namespace TreeSurvivalModel
{
    /// <summary>
    /// 树木生长相关算法接口(胸径、树高、干高、材积、叶面积年增量算法等);
    /// </summary>
    public interface IGrowthAlgorithm
    {
        double GrowthHeight2DBH(double dDBH, double dHmax, double dInitialSlopeDBHHeight);
        double GrowthDBH(double dCompetitionIndex, double dDEGDIndex, double dAridityIndex, double dTreeHeight, double dBoleHeight, double dDBH, double dGrowthHeight2DBH, double dLeafArea, double dSapwoodMaintenanceCostFactor, double dGrowthScalingFactor, double dPARResponseIntegral);
        double GrowthLeafArea(double dGrowthDBH, double dDBH, double
```

dLeafArea, double dSapwoodTurnoverRate, double dInitialLADDBH2Ratio);

  double GrowthTreeHeight(double dHmax, double dTreeHeight, double dGrowthDBH, double dDBH, double dInitialSlopeDBHHeight);

  double GrowthBoleHeight(double dK, double dBoleHeight, double dI0, double dLightCompensationPoint, double dHalfSaturationPoint, double dTreeHeight, double dLeafArea, double dCanopyArea, double dDeltaZ, string sLeafPhenology);

 }

 /// <summary>
 /// 树木死亡相关算法函数接口(生长效率、死亡概率算法等);
 /// </summary>
 public interface IMortalityAlgorithm
 {
  double TreeHeight(double dDBH, double dHmax, double ds);
  // double TotalSLA(double dDBH);
  int KnownSpeciesRadius(string sBaseSpeciesName, int iPositionValue);
  double TotalBA(DataTable dtNeighbors, string strFocalSpecies, int iFocalPosition, DataTable dtPhydist);

  double LAIofZOI(double dNeighborsSumLA, double dZOI);
  double NeighborsSumLA(DataTable dtNeighbors, string strFocalSpecies, int iFocalPosition, double dFocalTreeHeight, DataTable dtPhydist);
  double SpatialDistWeight(double dNeighborRadius, double dSpatialDistance);
  double PhyloDistWeight(double dPhyloDist);
  double NeighborRadius(string strSpecies, int iPosition, DataTable dtPhydist);
  double ZoneOfInfluence(double dRadius);
  double CompetitionIndex(double dTotalBA, double dZOI);
  double LeafAreaDensity(double dL, double dTreeHeight, double dBoleHeight, string sLeafPhenology, double dCanopyArea);

  double Fz(double dZ, double dSL, double dCanopyArea, double dTreeHeight, double dBoleHeight, string sLeafPhenology);
  double PARTreeTop(double dSLA, double dSubquadHillshade, double dRadius);// I(0);

```
        double PARCrown (double dI0, double dFz);// I(z);
        double PARResponse (double dHalfSaturationPoint, double
dLightCompensationPoint, double dIz); // P(z);
        double PARResponse0 (double dHalfSaturationPoint, double
dLightCompensationPoint, double dI0); // P(0);
        double PARResponseIntegral (double dTreeHeight, double dBoleHeight, double
dLeafArea, double dDeltaZ, double dHalfSaturationPoint, double dLightCompensationPoint,
double dI0, double dCanopyArea, string sPhenology); // P'

        double GrowthEfficiency (double dCompetitionIndex, double dDEGDIndex,
double dAridityIndex, double dTreeHeight, double dBoleHeight, double
dSapwoodMaintenanceCostFactor, double dGrowthScalingFactor, double dLeafArea,
double dDeltaZ, double dHalfSaturationPoint, double dLightCompensationPoint,
double dI0, double dCanopyArea, string sPhenology);
        //double IntrinsicMortalityRate (double dAgeMax);
        double DEGDResponse (double dSpeciesDEGDMax, double dSpeciesDEGDMin,
double dSiteDEGD);
        double AridityResponse (double dSiteAridity, double dSpeciesAridityMax);
        double IntrinsicMortalityRate (double dDBHmax, double dDBH, double
dIntrinsicMortalityRate);
        double WindthrowMortalityRate (string strRootType, double dSoilMoisture);
        double MortalityFunction (double dIntrinsicMortalityRate, double
dMortalityRateSuppression, double dWindthrowMortalityRate, double dRandomMortalityRate,
double dIndexOfVigour, double dThresholdVigor);
        double MortalityProbability (double dMortalityFunction, int iTimeStep);
        double HillshadeValueForIndividual (double dTreeX, double dTreeY);
        double SoilMoistureValueForIndividual (double dTreeX, double dTreeY);


    }

    /// <summary>
    /// 机理模型立地参数类；单件模式；
    /// </summary>
    public class SiteParameter
    {
```

附录 树木死亡动态模拟(TMDS)模型程序核心源代码(C#)

```csharp
private static SiteParameter uniqueSiteParameter;
private static readonly object _synLock = new object();
private double _AL00;
private double _k;
private double _DEGD;
private double _CCMax;
private double _aridity;
private double _deltaZ;
private double _dRho;
private double _dHillshadeMaxFDP;
private int _iYears;
private int _iTimeStep;
private DataTable dtPhydist;
private bool _bCanopyArchitecture;
private bool _bNeighborhoodRadius;
private bool _bPhyloCompetition;
private bool _bHabitatHeterogeneity;
private bool _bIntrinsicMortProb;
private bool _bExtrinsicMortProb;

private SiteParameter()
{
    _AL00 = 450;
    _k = 0.4;

    _DEGD = 1700;
    _CCMax = 60;
    _aridity = 0.74;
    _deltaZ = 3;
    _dRho = 999;
    _dHillshadeMaxFDP = 255;
    _iYears = 10;
    _iTimeStep = 1;
    dtPhydist = new DataTable("Phydist");
    _bCanopyArchitecture = true;
```

```
        _bNeighborhoodRadius = true;
        _bPhyloCompetition = true;
        _bHabitatHeterogeneity = true;
        _bIntrinsicMortProb = true;
        _bExtrinsicMortProb = true;
    }

    public static SiteParameter GetInstance ()
    {
        if (uniqueSiteParameter == null)
        {
            lock (_synLock)
            {
                if (uniqueSiteParameter == null)
                {
                    uniqueSiteParameter = new SiteParameter ();
                }
            }
        }
        return uniqueSiteParameter;
    }
    public double AL00
    {
        get { return _AL00; }
        set { _AL00 = value; }
    }
    public double K
    {
        get { return _k; }
        set { _k = value; }
    }
    public double DEGD
    {
        get { return _DEGD; }
```

```csharp
            set { _DEGD = value; }
        }
        public double CCMax
        {
            get { return _CCMax; }
            set { _CCMax = value; }
        }
        public double Aridity
        {
            get { return _aridity; }
            set { _aridity = value; }
        }
        public double DeltaZ
        {
            get { return _deltaZ; }
            set { _deltaZ = value; }
        }
        public double Rho
        {
            get { return _dRho; }
            set { _dRho = value; }
        }
        public double HillshadeMaxFDP
        {
            get { return _dHillshadeMaxFDP; }
            set { _dHillshadeMaxFDP = value; }
        }
        public int Years
        {
            get { return _iYears; }
            set { _iYears = value; }
        }
        public int TimeStep
        {
            get { return _iTimeStep; }
```

```csharp
            set { _iTimeStep = value; }
        }
        public DataTable Phylodist
        {
            get { return dtPhydist; }
            set { dtPhydist = value; }
        }
        public bool CanopyArchitecture
        {
            get { return _bCanopyArchitecture; }
            set { _bCanopyArchitecture = value; }
        }
        public bool NeighborhoodRadius
        {
            get { return _bNeighborhoodRadius; }
            set { _bNeighborhoodRadius = value; }
        }
        public bool PhyloCompetition
        {
            get { return _bPhyloCompetition; }
            set { _bPhyloCompetition = value; }
        }
        public bool HabitatHeterogeneity
        {
            get { return _bHabitatHeterogeneity; }
            set { _bHabitatHeterogeneity = value; }
        }
        public bool IntrinsicMortProb
        {
            get { return _bIntrinsicMortProb; }
            set { _bIntrinsicMortProb = value; }
        }
        public bool ExtrinsicMortProb
        {
            get { return _bExtrinsicMortProb; }
```

```csharp
            set { _bExtrinsicMortProb = value; }
        }
    }



    /// <summary>
    /// 树木死亡动态模拟抽象基类；
    /// </summary>
    public abstract class Model : IGrowthAlgorithm, IMortalityAlgorithm
    {
        protected SiteParameter siteParameter;
        protected DataTable dtSppParameters;
        protected Dataset dsHillshade;
        protected Dataset dsSoilMoisture;

        public Model(SiteParameter siteParameter, DataTable dtSppParameters, Dataset dsHillshade, Dataset dsSoilMoisture)
        {
            this.siteParameter = siteParameter;
            this.dtSppParameters = dtSppParameters;
            this.dtSppParameters.PrimaryKey = new DataColumn[] { dtSppParameters.Columns["Species"] };
            this.dsHillshade = dsHillshade;
            this.dsSoilMoisture = dsSoilMoisture;
        }
        public SiteParameter SiteParameters
        {
            get { return siteParameter; }
        }
        public DataTable SpeciesParameters
        {
            get { return dtSppParameters; }
        }
        public Dataset Hillshade
        {
```

```csharp
            get { return dsHillshade; }
        }
        public Dataset SoilMoisture
        {
            get { return dsSoilMoisture; }
        }

        public virtual double GrowthHeight2DBH(double dDBH, double dHmax, double dInitialSlopeDBHHeight) { return 0; }
        public virtual double GrowthDBH(double dCompetitionIndex, double dDEGDIndex, double dAridityIndex, double dTreeHeight, double dBoleHeight, double dDBH, double dGrowthHeight2DBH, double dLeafArea, double dSapwoodMaintenanceCostFactor, double dGrowthScalingFactor, double dPARResponseIntegral) { return 0; }
        public virtual double GrowthLeafArea(double dGrowthDBH, double dDBH, double dLeafArea, double dSapwoodTurnoverRate, double dInitialLADDBH2Ratio) { return 0; }
        public virtual double GrowthTreeHeight(double dHmax, double dTreeHeight, double dGrowthDBH, double dDBH, double dInitialSlopeDBHHeight) { return 0; }
        public virtual double GrowthBoleHeight(double dK, double dBoleHeight, double dI0, double dLightCompensationPoint, double dHalfSaturationPoint, double dTreeHeight, double dLeafArea, double dCanopyArea, double dDeltaZ, string sLeafPhenology) { return 0; }
        public virtual double TreeHeight(double dDBH, double dHmax, double ds) { return 0; }
        public virtual double DEGDResponse(double dSpeciesDEGDMax, double dSpeciesDEGDMin, double dSiteDEGD) { return 0; }
        public virtual double AridityResponse(double dSiteAridity, double dSpeciesAridityMax) { return 0; }

        public virtual int KnownSpeciesRadius(string sBaseSpeciesName, int iPositionValue)
        {
            int iNeignborRadius = 0;
            if (sBaseSpeciesName == "Pinus_koraiensis") //红松;
            {
```

```csharp
            if ((iPositionValue == 4) || (iPositionValue == 3))
            {
                iNeignborRadius = 6;
            }
            else if (iPositionValue == 2)
            {
                iNeignborRadius = 5;
            }
            else//1
            {
                iNeignborRadius = 5;
            }


        }
        else if ((sBaseSpeciesName == "Betula_costata") || (sBaseSpeciesName== "Betula_platyphylla"))
        {
            if ((iPositionValue == 4) || (iPositionValue == 3))
            {
                iNeignborRadius = 6;
            }
            else if (iPositionValue == 2)
            {
                iNeignborRadius = 5;
            }
            else//1
            {
                iNeignborRadius = 5;
            }
        }
        else if (sBaseSpeciesName == "Abies_nephrolepis")
        {
            if ((iPositionValue == 4) || (iPositionValue == 3))
            {
                iNeignborRadius = 5;
```

```
                }
                else if (iPositionValue == 2)
                {
                    iNeignborRadius = 7;
                }
                else//1
                {
                    iNeignborRadius = 6;
                }
            }
            else if (sBaseSpeciesName == "Picea_koraiensis" || sBaseSpeciesName == "Picea_jezoensis")
            {
                if ( ((iPositionValue == 4) || (iPositionValue == 3)) )
                {
                    iNeignborRadius = 7;
                }
                else if (iPositionValue == 2)
                {
                    iNeignborRadius = 6;
                }
                else//1
                {
                    iNeignborRadius = 6;
                }
            }
            else if (sBaseSpeciesName == "Juglans_mandshurica")
            {
                if ( ((iPositionValue == 4) || (iPositionValue == 3)) )
                {
                    iNeignborRadius = 5;
                }
                else if (iPositionValue == 2)
                {
                    iNeignborRadius = 5;
```

```csharp
            }
            else//1
            {
                iNeignborRadius = 5;
            }
        }
        else
        {
            if ( ( iPositionValue == 4) || ( iPositionValue == 3 ) )
            {
                iNeignborRadius = 6;
            }
            else if ( iPositionValue == 2)
            {
                iNeignborRadius = 5;
            }
            else//1
            {
                iNeignborRadius = 6;
            }
        }
        return iNeignborRadius;
    }
    public virtual double TotalBA (DataTable dtNeighbors, string strFocalSpecies, int iFocalPosition, DataTable dtPhydist) { return 0; }
    public virtual double LAIofZOI (double dNeighborsSumLA, double dZOI) { return 0; }
    public virtual double NeighborsSumLA (DataTable dtNeighbors, string strFocalSpecies, int iFocalPosition, double dFocalTreeHeight, DataTable dtPhydist) { return 0; }
    public virtual double SpatialDistWeight (double dNeighborRadius, double dSpatialDistance) { return 0; }
    public virtual double PhyloDistWeight (double dPhyloDist) { return 0; }
    public virtual double NeighborRadius (string sBaseSpeciesName, int iPositionValue, DataTable dtPhydist)
```

```
{
    IList<string> speciesList = new List<string>();
    speciesList.Add("Pinus_koraiensis");
    speciesList.Add("Betula_costata");
    speciesList.Add("Abies_nephrolepis");
    speciesList.Add("Picea_koraiensis");
    speciesList.Add("Picea_jezoensis");
    speciesList.Add("Betula_platyphylla");
    speciesList.Add("Juglans_mandshurica");
    //speciesList.Add("Juglans mandshurica Maxim");

    int iNeignborRadius = 0;
    if (speciesList.Contains(sBaseSpeciesName))
    {
        iNeignborRadius = KnownSpeciesRadius(sBaseSpeciesName, iPositionValue);
    }
    else
    {
        DataRow dr = dtPhydist.Rows.Find(sBaseSpeciesName);
        double dPhydist = 0;
        double dMinPhydist = 0;
        string sSimilarSpeices = "";
        dMinPhydist = 10000;
        foreach (string sSpecies in speciesList)
        {
            dPhydist = Convert.ToDouble(dr[sSpecies].ToString());
            if (dPhydist < dMinPhydist)
            {
                dMinPhydist = dPhydist;
                sSimilarSpeices = sSpecies;
            }
        }
        iNeignborRadius = KnownSpeciesRadius(sSimilarSpeices, iPositionValue);
```

```csharp
            }
            return iNeignborRadius;
        }
        public virtual double ZoneOfInfluence(double dRadius)
        {
            double dZOI = 0;
            return dZOI = Math.PI * Math.Pow(dRadius, 2);
        }
        public virtual double CompetitionIndex(double dTotalBA, double dZOI) { return 0; }
        public virtual double PARTreeTop(double dSLA, double dSubquadHillshade, double dRadius) { return 0; }
        public virtual double LeafAreaDensity(double dL, double dTreeHeight, double dBoleHeight, string sLeafPhenology, double dCanopyArea) { return 0; }
        public virtual double Fz(double dZ, double dSL, double dCanopyArea, double dTreeHeight, double dBoleHeight, string sLeafPhenology) { return 0; }
        public virtual double PARCrown(double dI0, double dFz) { return 0; }
        public virtual double PARResponse(double dHalfSaturationPoint, double dLightCompensationPoint, double dIz) { return 0; }
        public virtual double PARResponse0(double dHalfSaturationPoint, double dLightCompensationPoint, double dI0) { return 0; }
        public virtual double GrowthEfficiency(double dCompetitionIndex, double dDEGDIndex, double dAridityIndex, double dTreeHeight, double dBoleHeight, double dSapwoodMaintenanceCostFactor, double dGrowthScalingFactor, double dLeafArea, double dDeltaZ, double dHalfSaturationPoint, double dLightCompensationPoint, double dI0, double dCanopyArea, string sPhenology) { return 0; }
        public virtual double IntrinsicMortalityRate(double dDBHmax, double dDBH, double dIntrinsicMortalityRate) { return 0; }
        public virtual double MortalityFunction(double dIntrinsicMortalityRate, double dMortalityRateSuppression, double dWindthrowMortalityRate, double dRandomMortalityRate, double dIndexOfVigour, double dThresholdVigor) { return 0; }
        public virtual double MortalityProbability(double dMortalityFunction, int iTimeStep) { return 0; }
        public virtual double PARResponseIntegral(double dTreeHeight, double dBoleHeight, double dLeafArea, double dDeltaZ, double dHalfSaturationPoint, double
```

dLightCompensationPoint, double dI0, double dCanopyArea, string sPhenology)
{ return 0; }

  public virtual double WindthrowMortalityRate(string strRootType, double dSoilMoisture) { return 0; }

  public virtual double HillshadeValueForIndividual(double dTreeX, double dTreeY) { return 0; }

  public virtual double SoilMoistureValueForIndividual(double dTreeX, double dTreeY) { return 0; }

}

/// <summary>
/// TMDS 树木死亡动态模拟类；
/// /// </summary>
public class TMDS : Model
{
  public TMDS(SiteParameter siteParameter, DataTable dtSppParameters, Dataset dsHillshade, Dataset dsSoilMoisture) : base(siteParameter, dtSppParameters, dsHillshade, dsSoilMoisture) { }

  /// <summary>
  /// 树高相对胸径增长量；
  /// </summary>
  /// <param name="dDBH"></param>
  /// <param name="dHmax"></param>
  /// <param name="ds"></param>
  /// <returns></returns>
  public override double GrowthHeight2DBH(double dDBH, double dHmax, double dInitialSlopeDBHHeight)
  {
    double dGrowthHeight2DBH = 0;
    dGrowthHeight2DBH = dInitialSlopeDBHHeight * Math.Exp(-dInitialSlopeDBHHeight * dDBH / (dHmax - 1.3));
    return dGrowthHeight2DBH;

       }

       /// <summary>
       /// 胸径年增长量；
       /// </summary>
       /// <param name="dCompetitionIndex"></param>
       /// <param name="dDEGDIndex"></param>
       /// <param name="dAridityIndex"></param>
       /// <param name="dTreeHeight"></param>
       /// <param name="dBoleHeight"></param>
       /// <param name="dDBH"></param>
       /// <param name="dGrowthHeight2DBH"></param>
       /// <param name="dLeafArea"></param>
       /// <param name="dSapwoodMaintenanceCostFactor"></param>
       /// <param name="dGrowthScalingFactor"></param>
       /// <param name="dPARResponseIntegral"></param>
       /// <returns></returns>
       public override double GrowthDBH(double dCompetitionIndex, double dDEGDIndex, double dAridityIndex, double dTreeHeight, double dBoleHeight, double dDBH, double dGrowthHeight2DBH, double dLeafArea, double dSapwoodMaintenanceCostFactor, double dGrowthScalingFactor, double dPARResponseIntegral)
       {
              double dGrowthDBH = 0;
              double dD2Hdt = (dGrowthScalingFactor * dPARResponseIntegral - dSapwoodMaintenanceCostFactor * ((dTreeHeight + dBoleHeight) / 2)) * dLeafArea;

              double dStressFactor = 0;
              double dStressFactorTotal = 0;
              if (dDEGDIndex >= dAridityIndex)
              {
                     dStressFactor = dAridityIndex;
              }
              else
              {

```
                    dStressFactor = dDEGDIndex;
                }
                if (dCompetitionIndex >= dStressFactor)
                {
                    dStressFactorTotal = dStressFactor;
                }
                else
                {
                    dStressFactorTotal = dCompetitionIndex;
                }
                dGrowthDBH = dStressFactorTotal * (dGrowthScalingFactor *
dPARResponseIntegral - dSapwoodMaintenanceCostFactor * ((dTreeHeight +
dBoleHeight) / 2)) * dLeafArea / ((2 * dTreeHeight / dDBH + dGrowthHeight2DBH)
* Math.Pow(dDBH, 2));

                if (dGrowthDBH < 0)
                {
                    dGrowthDBH = 0;
                }
                return dGrowthDBH;
            }


            /// <summary>
            /// 叶面积年增长量;
            /// </summary>
            /// <param name="dGrowthDBH"></param>
            /// <param name="dDBH"></param>
            /// <param name="dLeafArea"></param>
            /// <param name="dSapwoodTurnoverRate"></param>
            /// <param name="dInitialLADDBH2Ratio"></param>
            /// <returns></returns>
            public override double GrowthLeafArea(double dGrowthDBH, double dDBH,
double dLeafArea, double dSapwoodTurnoverRate, double dInitialLADDBH2Ratio)
            {
                double dGrowthLeafArea = 0;
```

```csharp
            dGrowthLeafArea = dGrowthDBH * 2 * dInitialLADDBH2Ratio * dDBH - dSapwoodTurnoverRate * dLeafArea;
            return dGrowthLeafArea;
        }


        /// <summary>
        /// 干高年增长量;
        /// </summary>
        /// <param name="dK"></param>
        /// <param name="dNewLeafArea"></param>
        /// <param name="dBoleHeight"></param>
        /// <param name="dNewTreeHeight"></param>
        /// <param name="dI0"></param>
        /// <param name="dLightCompensationPoint"></param>
        /// <returns></returns>
        public override double GrowthBoleHeight(double dK, double dBoleHeight, double dI0, double dLightCompensationPoint, double dHalfSaturationPoint, double dTreeHeight, double dLeafArea, double dCanopyArea, double dDeltaZ, string sLeafPhenology)
        {
            double dGrowthBoleHeight = 0;
            int iCount = Convert.ToInt32((dTreeHeight - dBoleHeight) / siteParameter.DeltaZ);
            int iRealCount = 0;
            for (int i = 1; i < iCount; i++)
            {
                iRealCount = i;
                double dZ = dDeltaZ * i;
                double dFz = Fz(dZ, LeafAreaDensity(dLeafArea, dTreeHeight, dBoleHeight, sLeafPhenology, dCanopyArea), dCanopyArea, dTreeHeight, dBoleHeight, sLeafPhenology);
                double dIz = PARCrown(dI0, dFz);
                if (dIz <= dLightCompensationPoint)
                {
```

```
                iRealCount = i;
                break;
            }

        dGrowthBoleHeight = dTreeHeight - dBoleHeight - iRealCount * dDeltaZ;
        if (dGrowthBoleHeight < 0)
        {
            int iNegative = 0;
        }
        // dGrowthBoleHeight = -Math.Log(Math.E, dLightCompensationPoint /
dK * dI0) / dK * LeafAreaDensity(dNewLeafArea, dNewTreeHeight, dBoleHeight);
        return dGrowthBoleHeight;
    }



    /// <summary>
    /// 树高年增长量;
    /// </summary>
    /// <param name="dHmax"></param>
    /// <param name="dTreeHeight"></param>
    /// <param name="dGrowthDBH"></param>
    /// <param name="dDBH"></param>
    /// <param name="dInitialSlopeDBHHeight"></param>
    /// <returns></returns>
    public override double GrowthTreeHeight(double dHmax, double dTreeHeight,
double dGrowthDBH, double dDBH, double dInitialSlopeDBHHeight)
    {
        double dGrowthTreeHeight = 0;
        double dNewDBH = dDBH + dGrowthDBH;
        double dTreeHeight0 = TreeHeight(dDBH, dHmax,
dInitialSlopeDBHHeight);
        double dNewTreeHeight = TreeHeight(dNewDBH, dHmax,
dInitialSlopeDBHHeight);
        dGrowthTreeHeight = dNewTreeHeight - dTreeHeight0;
        if (dGrowthTreeHeight < 0)
```

```
            {
                int iNegative = 0;
            }
            return dGrowthTreeHeight;
        }

        /// <summary>
        /// 积温响应函数;
        /// </summary>
        /// <param name="dSpeciesDEGDMax"></param>
        /// <param name="dSpeciesDEGDMin"></param>
        /// <param name="dSiteDEGD"></param>
        /// <returns></returns>
        public override double DEGDResponse(double dSpeciesDEGDMax, double dSpeciesDEGDMin, double dSiteDEGD)
        {
            double dDEGDIndex = 0;
            dDEGDIndex = 4 * (dSpeciesDEGDMax - dSiteDEGD) * (dSiteDEGD - dSpeciesDEGDMin) / Math.Pow((dSpeciesDEGDMax - dSpeciesDEGDMin), 2);
            if (dDEGDIndex > 0)
            {
                return dDEGDIndex;
            }
            else
            {
                return 0;
            }
        }
        /// <summary>
        /// 干燥度响应函数;
        /// </summary>
        /// <param name="dSiteAridity"></param>
        /// <param name="dSpeciesAridityMax"></param>
        /// <returns></returns>
```

```csharp
public override double AridityResponse(double dSiteAridity, double dSpeciesAridityMax)
{
    double dAridityIndex = 0;
    dAridityIndex = 4 * dSiteAridity * (dSpeciesAridityMax - dSiteAridity) / Math.Pow(dSpeciesAridityMax, 2);
    return dAridityIndex;
}

/// <summary>
/// 树高曲线方程;
/// </summary>
/// <param name="dDBH"></param>
/// <param name="dHmax"></param>
/// <param name="ds"></param>
/// <returns></returns>
public override double TreeHeight(double dDBH, double dHmax, double ds)
{
    double dTreeHeight = 0;
    return dTreeHeight = 1.3 + (dHmax - 1.3) * (1 - Math.Exp(-ds * dDBH / (dHmax - 1.3)));
}

/// <summary>
/// ZOI 区域内叶面积指数;
/// </summary>
/// <param name="dNeighborsSumLA"></param>
/// <param name="dZOI"></param>
/// <returns></returns>
public override double LAIofZOI(double dNeighborsSumLA, double dZOI)
{
    double dLAIofZOI = 0;
    dLAIofZOI = dNeighborsSumLA / dZOI;
    return dLAIofZOI;
}
```

/// <summary>
/// 高于基株的邻体叶面积加权和;
/// </summary>
/// <param name="dtNeighbors"></param>
/// <param name="strFocalSpecies"></param>
/// <param name="iFocalPosition"></param>
/// <param name="dFocalTreeHeight"></param>
/// <param name="dtPhydist"></param>
/// <returns></returns>
```csharp
public override double NeighborsSumLA(DataTable dtNeighbors, string strFocalSpecies, int iFocalPosition, double dFocalTreeHeight, DataTable dtPhydist)
{
    double dSurvProbability = 0;
    double dNeighborsSLA = 0;
    double dSpatialDistance = 0;
    double dPhyloDistance = 0;
    double dNeighborWeightSLA = 0;
    double dSumNeighborWeightSLA = 0;
    double dNeighborTreeHeight = 0;
    double dRadius = 0;
    if (siteParameter.NeighborhoodRadius == true)
    {
        dRadius=NeighborRadius(strFocalSpecies, iFocalPosition, dtPhydist);
    }
    else
    {
        dRadius = 7;
    }

    foreach (DataRow neighbor in dtNeighbors.Rows)
    {
        dSurvProbability = Convert.ToDouble(neighbor["SurvProb"]);
        dNeighborTreeHeight = Convert.ToDouble(neighbor["TreeHeight"]);
```

```
long tick5 = DateTime.Now.Ticks;
Random random5 = new Random((int)(tick5 & 0xffffffffL) | (int)(tick5 >> 32));
int iRandomNumber5 = random5.Next(50, 100);
double dRandomMortalityRate5 = iRandomNumber5 * 0.01;


if (dSurvProbability > dRandomMortalityRate5)
{
    if (dNeighborTreeHeight > dFocalTreeHeight)
    {
        dNeighborsSLA = Convert.ToDouble(neighbor["LeafArea"]);
        dSpatialDistance = Convert.ToDouble(neighbor["SpatialDistance"]);
        dPhyloDistance = Convert.ToDouble(neighbor["PhyloDistance"]);

        if (siteParameter.PhyloCompetition == true)
        {
            dNeighborWeightSLA = dNeighborsSLA * SpatialDistWeight(dRadius, dSpatialDistance) * PhyloDistWeight(dPhyloDistance);
        }
        else
        {
            dNeighborWeightSLA = dNeighborsSLA * SpatialDistWeight(dRadius, dSpatialDistance);
        }

        //dNeighborWeightSLA = dNeighborsSLA * SpatialDistWeight(dRadius, dSpatialDistance) * PhyloDistWeight(dPhyloDistance);
        //dNeighborWeightSLA = dNeighborsSLA * SpatialDistWeight(dRadius, dSpatialDistance);
        dSumNeighborWeightSLA = dNeighborWeightSLA + dSumNeighborWeightSLA;
    }
```

```csharp
            }
        }
        return dSumNeighborWeightSLA;
    }


    /// <summary>
    /// 竞争乘数因子函数;
    /// </summary>
    /// <param name="dTotalBA"></param>
    /// <param name="dZOI"></param>
    /// <returns></returns>
    public override double CompetitionIndex(double dTotalBA, double dZOI)
    {
        double dCompetitionIndex = 0;
        return dCompetitionIndex = 1-dTotalBA/(siteParameter.CCMax * dZOI);
    }


    /// <summary>

    /// </summary>
    /// <param name="dSLA"></param>
    /// <param name="dSubquadHillshade"></param>
    /// <param name="dRadius"></param>
    /// <returns></returns>
    public override double PARTreeTop(double dSLA, double dSubquadHillshade, double dRadius)
    {
        double dI0 = 0;
        double dAL0 = 0;

        dAL0 = siteParameter.AL00 * dSubquadHillshade / siteParameter.HillshadeMaxFDP;
        dAL0 = siteParameter.AL00 * dSubquadHillshade / 179;
        dAL0 = siteParameter.AL00 * dSubquadHillshade / 196; //Median
```

```
        double dNeighbRadius = 0;

        if (siteParameter.NeighborhoodRadius == true)
        {
            dNeighbRadius = dRadius;
        }
        else
        {
            dNeighbRadius = 7;
        }

        double dZOI = ZoneOfInfluence(dNeighbRadius);
        double dLAIofZOI = LAIofZOI(dSLA, dZOI);
        return dI0 = dAL0 * Math.Exp(-siteParameter.K * dLAIofZOI);
    }

    /// <summary>
    /// 叶面积密度函数;
    /// </summary>
    /// <param name="dLeafArea"></param>
    /// <param name="dHeight"></param>
    /// <param name="dCH"></param>
    /// <returns></returns>
    public override double LeafAreaDensity(double dLeafArea, double dTreeHeight, double dBoleHeight, string sLeafPhenology, double dCanopyArea)
    {
        if (siteParameter.CanopyArchitecture == true)
        {
            double dSL = 0;
            double d2R = 2 * Math.Sqrt(dCanopyArea / Math.PI);
```

```csharp
            double dA = Math.Sqrt(dCanopyArea / Math.PI);
            double dB = Math.Sqrt(dCanopyArea / Math.PI);
            double dC = (dTreeHeight - dBoleHeight) / 2;
            return dSL = dLeafArea / ((4 / 3) * Math.PI * dA * dB * dC);
        }
        else
        {
            double dSL;
            return dSL = dLeafArea / (dTreeHeight - dBoleHeight);
        }
    }

    /// <summary>
    /// 光合作用响应函数; P(z); 0-1;
    /// </summary>
    /// <param name="K"></param>
    /// <param name="HalfSaturationPoint"></param>
    /// <param name="LightCompensationPoint"></param>
    /// <param name="dIz"></param>
    /// <returns></returns>
    public override double PARResponse(double dHalfSaturationPoint, double dLightCompensationPoint, double dIz)
    {
        double dPz;
        double a = siteParameter.K * dIz - dLightCompensationPoint;
        double b = dHalfSaturationPoint - dLightCompensationPoint;

        dPz = (siteParameter.K * dIz - dLightCompensationPoint) / (dHalfSaturationPoint - dLightCompensationPoint);
        if (dPz > 1)
        {
            dPz = 1;
        }
        if (dPz < 0)
        {
```

```
            dPz = 0;
        }

        return dPz;
    }

    /// <summary>
    /// 树梢光合作用响应函数; P(0);
    /// </summary>
    /// <param name="K"></param>
    /// <param name="HalfSaturationPoint"></param>
    /// <param name="LightCompensationPoint"></param>
    /// <param name="dI0"></param>
    /// <returns></returns>
    public override double PARResponse0(double dHalfSaturationPoint, double dLightCompensationPoint, double dI0)
    {
        double dPz;

        dPz = (siteParameter.K * dI0 - dLightCompensationPoint) / (dHalfSaturationPoint - dLightCompensationPoint);
        if (dPz > 1)
        {
            dPz = 1;
        }
        else if (dPz < 0)
        {
            dPz = 0;
        }
        return dPz;
        //return dPz = (siteParameter.K * dI0 - dLightCompensationPoint) / (siteParameter.K * dI0 + dHalfSaturationPoint - dLightCompensationPoint);
    }

    /// <summary>
```

/// 树梢向下 Zm 处光合有效辐射函数；
/// </summary>
/// <param name="dI0"></param>
/// <param name="K"></param>
/// <param name="dFz"></param>
/// <returns></returns>
public override double PARCrown(double dI0, double dFz)
{
    double dIz;
    return dIz = dI0 * Math.Exp(-siteParameter.K * dFz);
}

/// <summary>
/// 树梢至 Zm 处的叶面积指数函数；
/// </summary>
/// <param name="dZ"></param>
/// <param name="dSL"></param>
/// <returns></returns>
public override double Fz(double dZ, double dSL, double dCanopyArea, double dTreeHeight, double dBoleHeight, string sLeafPhenology)
{

    if (siteParameter.CanopyArchitecture == true)
    {
        double dFz = 0;

        double dR = Math.Sqrt(dCanopyArea / Math.PI);
        double dVolume = 0;

        double dA = dR;
        double dB = dR;
        double dC = (dTreeHeight - dBoleHeight) / 2;
        double dH = dTreeHeight - dBoleHeight - dZ;

        if (dVolume < 0)

```
                {
                    int iV = 11;
                }
                double dCurrentCanopyArea = 0;
                if (dZ > dC)
                {
                    double dTotalVolume = (4 / 3) * Math.PI * dA * dB * dC;
                    double dRestVolume = 0;
                    dH = dH - dC;
                    dRestVolume = Math.PI * dA * dB * ((2 / 3) * dC - dH + dH
* dH / (3 * dC * dC));
                    dVolume = dTotalVolume - dRestVolume;
                    dCurrentCanopyArea = dCanopyArea;
                    dFz = (dSL * dVolume) / dCurrentCanopyArea;
                }
                else
                {
                    dVolume = Math.PI * dA * dB * ((2 / 3) * dC - dH + dH *
dH / (3 * dC * dC));
                    dCurrentCanopyArea = 3 * dVolume / dZ;
                    dFz = (dSL * dVolume) / dCurrentCanopyArea;

                }

                return dFz;
            }
            else
            {
                double dFz;
                return dFz = (dSL * dZ) / dCanopyArea;
            }

        }

        /// <summary>
```

```
/// 光合作用函数定积分;
/// </summary>
/// <param name="dTreeHeight"></param>
/// <param name="dBoleHeight"></param>
/// <param name="dLeafArea"></param>
/// <param name="dDeltaZ"></param>
/// <param name="dHalfSaturationPoint"></param>
/// <param name="dLightCompensationPoint"></param>
/// <param name="dI0"></param>
/// <returns></returns>
public override double PARResponseIntegral(double dTreeHeight, double dBoleHeight, double dLeafArea, double dDeltaZ, double dHalfSaturationPoint, double dLightCompensationPoint, double dI0, double dCanopyArea, string sPhenology)
{
    double dPARResponseIntegral = 0;
    int iCount = Convert.ToInt32((dTreeHeight - dBoleHeight) / siteParameter.DeltaZ);
    int iRealCount = 1;

    for (int i = 1; i < iCount; i++)
    {
        iRealCount = i;
        double dZ = dDeltaZ * i;
        double dFz = Fz(dZ, LeafAreaDensity(dLeafArea, dTreeHeight, dBoleHeight, sPhenology, dCanopyArea), dCanopyArea, dTreeHeight, dBoleHeight, sPhenology);
        double dIz = PARCrown(dI0, dFz);
        if (dIz <= dLightCompensationPoint)
        {
            iRealCount = i;
            break;
        }
        double dPz = PARResponse(dHalfSaturationPoint, dLightCompensationPoint, dIz);
        dPARResponseIntegral = dPARResponseIntegral + dPz;
```

}


```
/// <summary>
/// 生长效率函数；生长效率是实际生长速率与最大生长速率的比值；
/// </summary>
/// <param name="dCompetitionIndex"></param>
/// <param name="dDEGDIndex"></param>
/// <param name="dAridityIndex"></param>
/// <param name="dTreeHeight"></param>
/// <param name="dBoleHeight"></param>
/// <param name="dSapwoodMaintenanceCostFactor"></param>
/// <param name="dGrowthScalingFactor"></param>
/// <param name="dLeafArea"></param>
/// <param name="dDeltaZ"></param>
/// <param name="dHalfSaturationPoint"></param>
/// <param name="dLightCompensationPoint"></param>
/// <param name="dI0"></param>
/// <returns></returns>
public override double GrowthEfficiency(double dCompetitionIndex, double dDEGDIndex, double dAridityIndex, double dTreeHeight, double dBoleHeight, double dSapwoodMaintenanceCostFactor, double dGrowthScalingFactor, double dLeafArea, double dDeltaZ, double dHalfSaturationPoint, double dLightCompensationPoint, double dI0, double dCanopyArea, string sPhenology)
{
    double dGrowthEfficiency = 0;
    double dEmax = 0;
    double dErea = 0;
    double dPARResponse0 = PARResponse0(dHalfSaturationPoint, dLightCompensationPoint, dI0);
    double dPARResponseIntegral = PARResponseIntegral(dTreeHeight, dBoleHeight, dLeafArea, dDeltaZ, dHalfSaturationPoint, dLightCompensationPoint, dI0, dCanopyArea, sPhenology);
    dEmax = dGrowthScalingFactor * PARResponse0(dHalfSaturationPoint,
```

```csharp
dLightCompensationPoint, dI0) * dLeafArea;

            double dStressFactor = 0;
            double dStressFactorTotal = 0;
            if (dDEGDIndex >= dAridityIndex)
            {
                dStressFactor = dAridityIndex;
            }
            else
            {
                dStressFactor = dDEGDIndex;
            }
            if (dCompetitionIndex >= dStressFactor)
            {
                dStressFactorTotal = dStressFactor;
            }
            else
            {
                dStressFactorTotal = dCompetitionIndex;
            }
            dErea = dStressFactorTotal * (dGrowthScalingFactor *
PARResponseIntegral(dTreeHeight, dBoleHeight, dLeafArea, dDeltaZ, dHalfSaturationPoint,
dLightCompensationPoint, dI0, dCanopyArea, sPhenology) -
dSapwoodMaintenanceCostFactor * ((dTreeHeight - dBoleHeight) / 2)) * dLeafArea;


            dGrowthEfficiency = dErea / dEmax;

            if (dGrowthEfficiency < 0)
            {
                dGrowthEfficiency = 0.000001;
            }

            return dGrowthEfficiency;
        }
```

```csharp
/// <summary>
/// 基株周围邻体树木总胸高断面积；
/// </summary>
/// <param name="dtNeighbors"></param>
/// <returns></returns>
public override double TotalBA (DataTable dtNeighbors, string strFocalSpecies, int iFocalPosition, DataTable dtPhydist)
{
    double dSurvProbability = 0;
    double dNeighborBA = 0;
    double dSpatialDistance = 0;
    double dPhyloDistance = 0;
    double dNeighborWeightBA = 0;
    double dSumNeighborWeightBA = 0;
    double dRadius = 0;

    if (siteParameter.NeighborhoodRadius == true)
    {
        dRadius = NeighborRadius (strFocalSpecies, iFocalPosition, dtPhydist);
    }
    else
    {
        dRadius = 7;
    }

    foreach (DataRow neighbor in dtNeighbors.Rows)
    {
        dSurvProbability = Convert.ToDouble (neighbor["SurvProb"]);


        long tick6 = DateTime.Now.Ticks;
        Random random6 = new Random ((int) (tick6 & 0xffffffffL) | (int) (tick6 >> 32));
        int iRandomNumber6 = random6.Next (50, 100);
```

```csharp
                double dRandomMortalityRate6 = iRandomNumber6 * 0.01;

                if (dSurvProbability > dRandomMortalityRate6)
                {

                    dNeighborBA = Convert.ToDouble(neighbor["BA"]);
                    dSpatialDistance= Convert.ToDouble(neighbor["SpatialDistance"]);
                    dPhyloDistance = Convert.ToDouble(neighbor["PhyloDistance"]);

                    if (siteParameter.PhyloCompetition == true)
                    {
                        dNeighborWeightBA = dNeighborBA * SpatialDistWeight(dRadius, dSpatialDistance) * PhyloDistWeight(dPhyloDistance);
                    }
                    else
                    {
                        dNeighborWeightBA = dNeighborBA * SpatialDistWeight(dRadius, dSpatialDistance);
                    }
                    //dNeighborWeightBA = dNeighborBA * SpatialDistWeight(dRadius, dSpatialDistance) * PhyloDistWeight(dPhyloDistance);

                    dSumNeighborWeightBA = dNeighborWeightBA + dSumNeighborWeightBA;
                }

                dNeighborBA = Convert.ToDouble(neighbor["BA"]);
            }
            return dSumNeighborWeightBA;
        }

        /// <summary>
        /// 系统发育距离(基株与邻体间亲缘关系)权重函数;
        /// </summary>
        /// <param name="dPhyloDist"></param>
```

/// <returns></returns>
public override double PhyloDistWeight(double dPhyloDist)
{
    double dPhyloDistWeight = 0;
    double dNugget = 0;
    double dStill = 0.4;
    double dPracticalRange = 600;
    double dPhyloDistWeight1;
    dPhyloDistWeight1 = dNugget + dStill * (1 - Math.Exp(-3 * dPhyloDist / dPracticalRange));
    dPhyloDistWeight = 1 - dPhyloDistWeight1;
    return dPhyloDistWeight;
}

/// <summary>
/// 空间距离权重函数；
/// </summary>
/// <param name="dNeighborRadius"></param>
/// <param name="dSpatialDistance"></param>
/// <returns></returns>
public override double SpatialDistWeight(double dNeighborRadius, double dSpatialDistance)
{
    double dSpatialDistWeight = 0;
    double dTHE = 0;
    dTHE=2*Math.Acos(dSpatialDistance/Convert.ToInt32(dNeighborRadius));
    dSpatialDistWeight = (dTHE - Math.Sin(dTHE)) / Math.PI;
    return dSpatialDistWeight;
}

/// <summary>
/// 内禀死亡率；
/// </summary>
/// <param name="dDBHmax"></param>

/// <param name="dDBH"></param>
/// <returns></returns>
public override double IntrinsicMortalityRate(double dDBHmax, double dDBH, double dIntrinsicMortalityRate)
{
    double dU0 = 0;
    double dU00 = 0;
    dU0 = dIntrinsicMortalityRate;
    if (siteParameter.IntrinsicMortProb == true)
    {
        double dK = 0.1;
        dU00 = dU0 * (1 - dK * Math.Sin(Math.PI * dDBH / dDBHmax));
        return dU00;
    }
    else
    {
        return dU0;
    }
}

/// <summary>
/// 死亡函数;
/// </summary>
/// <param name="dIntrinsicMortalityRate"></param>
/// <param name="dMortalityRateSuppression"></param>
/// <param name="dIndexOfVigour"></param>
/// <param name="dThresholdVigor"></param>
/// <returns></returns>
public override double MortalityFunction(double dIntrinsicMortalityRate, double dMortalityRateSuppression, double dWindthrowMortalityRate, double dRandomMortalityRate, double dIndexOfVigour, double dThresholdVigor)
{
    double dMortalityFunction = 0;
    if (siteParameter.ExtrinsicMortProb == false)
    {

```
            dMortalityRateSuppression = dMortalityRateSuppression * 5;
      }

            return dMortalityFunction = dIntrinsicMortalityRate +
dWindthrowMortalityRate + dRandomMortalityRate + dMortalityRateSuppression / (1 +
Math.Pow((dIndexOfVigour / dThresholdVigor), siteParameter.Rho));
      }

      /// <summary>
      /// 死亡概率函数;
      /// </summary>
      /// <param name="dMortalityFunction"></param>
      /// <param name="iYearInterval"></param>
      /// <returns></returns>
      public override double MortalityProbability(double dMortalityFunction, int iTimeStep)
      {
            double dMortalityProbability = 0;
            return dMortalityProbability=1-Math.Exp(-dMortalityFunction* iTimeStep);

      }
      /// <summary>
      /// 风倒木死亡率; 风倒木死亡率由根系深浅以及土壤湿度共同决定;
      /// </summary>
      /// <param name="strRootType"></param>
      /// <param name="dSoilMoisture"></param>
      /// <returns></returns>
      public override double WindthrowMortalityRate(string strRootType, double dSoilMoisture)
      {
            double dWindthrowMortalityRate = 0;
            double dPracticalRange = 50;
            double dStill = 0.03;
            if (strRootType == "Shallow")
            {
```

```csharp
            dWindthrowMortalityRate = dStill * (1-Math.Exp(-3 * dSoilMoisture / dPracticalRange));
                return dWindthrowMortalityRate;
            }
            else //Deep-rootedness
            {
                return 0;
            }

        }
        /// <summary>
        /// 根据树木坐标提取 Hillshade 值；
        /// </summary>
        /// <param name="dTreeX"></param>
        /// <param name="dTreeY"></param>
        /// <returns></returns>
        public override double HillshadeValueForIndividual(double dTreeX, double dTreeY)
        {
            double dHillshadeValue = 0;
            Band bandHillshade = dsHillshade.GetRasterBand(1);
            double[] dGeotransform = new double[6];
            dsHillshade.GetGeoTransform(dGeotransform);
            double dXCellSize = dGeotransform[1];
            double dYCellSize = dGeotransform[5];
            double[] dBuf = new double[1];
            double dRasterX = (dTreeX - dGeotransform[0]) / dGeotransform[1];
            double dRasterY = (dTreeY - dGeotransform[3]) / dGeotransform[5];
            int iRasterX = Convert.ToInt32(dRasterX);
            if (iRasterX >= 60)
            {
                iRasterX = 59;
            }
            int iRasterY = Convert.ToInt32(dRasterY);
```

```
            if (iRasterY >= 60)
            {
                iRasterY = 59;
            }
            bandHillshade.ReadRaster(iRasterX, iRasterY, 1, 1, dBuf, 1, 1, 0, 0);
            dHillshadeValue = dBuf[0];
            if ((dHillshadeValue == 0) && (dTreeX > 280) && (dTreeX <= 295))
            {
                bandHillshade.ReadRaster(iRasterX-1, iRasterY, 1, 1, dBuf, 1, 1, 0, 0);
                dHillshadeValue = dBuf[0];
            }
            else if ((dHillshadeValue == 0) && (dTreeY < 30) && (dTreeY >= 5))
            {
                bandHillshade.ReadRaster(iRasterX, iRasterY-1, 1, 1, dBuf, 1, 1, 0, 0);
                dHillshadeValue = dBuf[0];
            }
            return dHillshadeValue;
        }


        /// <summary>
        /// 根据树木坐标提取土壤湿度值;
        /// </summary>
        /// <param name="dTreeX"></param>
        /// <param name="dTreeY"></param>
        /// <returns></returns>
        public override double SoilMoistureValueForIndividual(double dTreeX, double dTreeY)
        {
            double dSoilMoistureValue = 0;
            Band bandSoilMoisture = dsSoilMoisture.GetRasterBand(1);
            double[] dGeotransform = new double[6];
            dsSoilMoisture.GetGeoTransform(dGeotransform);
            double dXCellSize = dGeotransform[1];
            double dYCellSize = dGeotransform[5];
            double[] dBuf = new double[1];
```

```csharp
            double dRasterX = (dTreeX - dGeotransform[0]) / dGeotransform[1];
            double dRasterY = (dTreeY - dGeotransform[3]) / dGeotransform[5];
            int iRasterX = Convert.ToInt32(dRasterX);
            if (iRasterX >= 60)
            {
                iRasterX = 59;
            }
            int iRasterY = Convert.ToInt32(dRasterY);
            if (iRasterY >= 60)
            {
                iRasterY = 59;
            }
            bandSoilMoisture.ReadRaster(iRasterX, iRasterY, 1, 1, dBuf, 1, 1, 0, 0);
            dSoilMoistureValue = dBuf[0];
            if ((dSoilMoistureValue == 0) && (dTreeX > 280) && (dTreeX <= 295))
            {
                bandSoilMoisture.ReadRaster(iRasterX - 1, iRasterY, 1, 1, dBuf, 1, 1, 0, 0);
                dSoilMoistureValue = dBuf[0];
            }
            else if ((dSoilMoistureValue == 0) && (dTreeY < 30) && (dTreeY >= 5))
            {
                bandSoilMoisture.ReadRaster(iRasterX, iRasterY - 1, 1, 1, dBuf, 1, 1, 0, 0);
                dSoilMoistureValue = dBuf[0];
            }
            return dSoilMoistureValue;
        }

    }

    /// <summary>
    /// 树木抽象基类；模板方法模式；
    /// ODD protocal;
    /// </summary>
```

```
public abstract class Tree
{
    protected Model model;
    protected FeatureEx featureEx;
    protected DataTable dtNeighbors;
    protected Dictionary<string, double> growthDictionary;
    protected double dTreeX, dTreeY;
    protected double dDBH;
    protected double dBA;
    protected string sSpecies;
    protected double dTreeHeight;
    protected double dBoleHeight;
    protected double dLAI;
    protected double dLeafArea;
    protected double dCanopyArea;
    protected int iiTopographicPosition;
    protected double dHmax;
    protected double dInitialSlopeDBHHeight;
    protected double dHalfSaturationPoint;
    protected double dLightCompensationPoint;
    protected double dIntrinsicMortalityRate;
    protected double dWindthrowMortalityRate;
    protected double dMortalityRateSuppression;
    protected double dThresholdVigor;
    protected double dSapwoodMaintenanceCostFactor;
    protected double dGrowthScalingFactor;
    protected double dDEGDMax, dDEGDMin;
    protected double dAridityMax;
    protected double dSapwoodTurnoverRate;
    protected double dInitialLADDBH2Ratio;
    protected string sPhenology;
    protected double dDBHmax;
    protected string strRootType;

    public Tree(Model model, Feature feature)
```

```csharp
        {
            this.model = model;
            featureEx = new FeatureEx(feature);
            //this.feature = feature;
        }

        private void CommonAlgorithm()
        {
            Gdal.SetConfigOption("SHAPE_ENCODING", "");
            Gdal.SetConfigOption("GDAL_FILENAME_IS_UTF8", "YES");
            Ogr.RegisterAll();
            Gdal.AllRegister();
            dTreeX = Convert.ToDouble(featureEx.GetFieldValue("gx"));
            dTreeY = Convert.ToDouble(featureEx.GetFieldValue("gy"));
            dDBH = Convert.ToDouble(featureEx.GetFieldValue("DBH"));
            dBA = Convert.ToDouble(featureEx.GetFieldValue("BA"));
            sSpecies = Convert.ToString(featureEx.GetFieldValue("Species"));
            dTreeHeight= Convert.ToDouble(featureEx.GetFieldValue("TreeHeight"));
            dBoleHeight= Convert.ToDouble(featureEx.GetFieldValue("BoleHeight"));
            dLeafArea = Convert.ToDouble(featureEx.GetFieldValue("LeafArea"));
            dCanopyArea=Convert.ToDouble(featureEx.GetFieldValue("CanopyArea"));
            dLAI = Convert.ToDouble(featureEx.GetFieldValue("LAI"));
            iiTopographicPosition = Convert.ToInt32(featureEx.GetFieldValue("Position"));

            DataRow dr = model.SpeciesParameters.Rows.Find(sSpecies);
            dHmax = Convert.ToDouble(dr["Hmax"].ToString());
            dInitialSlopeDBHHeight =Convert.ToDouble(dr["SlopeDH"].ToString());

            dDBHmax = Convert.ToDouble(dr["DBHmax"].ToString());
            dHalfSaturationPoint = Convert.ToDouble(dr["Saturation"].ToString());
            dLightCompensationPoint = Convert.ToDouble(dr["Compensati"].ToString());
            dIntrinsicMortalityRate =Convert.ToDouble(dr["IntrinsicM"].ToString());
            dMortalityRateSuppression =Convert.ToDouble(dr["Suppressio"].ToString());
            dThresholdVigor = Convert.ToDouble(dr["ThresholdV"].ToString());
```

```
        dSapwoodMaintenanceCostFactor =
Convert.ToDouble(dr["SapwMaint"].ToString());
        dGrowthScalingFactor = Convert.ToDouble(dr["GrowScale"].ToString());
        dDEGDMax = Convert.ToDouble(dr["DEGDMax"].ToString());
        dDEGDMin = Convert.ToDouble(dr["DEGDMin"].ToString());
        dAridityMax = Convert.ToDouble(dr["AridityMax"].ToString());
        dSapwoodTurnoverRate = Convert.ToDouble(dr["SapwTurn"].ToString());
        dInitialLADDBH2Ratio =Convert.ToDouble(dr["LAD2Ratio"].ToString());
        sPhenology = Convert.ToString(dr["Leaf_pheno"].ToString());
        strRootType = Convert.ToString(dr["RootType"].ToString());
    }
    public abstract double MortalityAlgorithm();
    public abstract Dictionary<string, double> GrowthAlgorithm();
    public Dictionary<string, double> Growth()
    {
        growthDictionary = new Dictionary<string, double>();
        CommonAlgorithm();
        growthDictionary = GrowthAlgorithm();
        return growthDictionary;
    }
    public double Mortality()
    {
        double mortalityResult = 0;
        CommonAlgorithm();
        mortalityResult = MortalityAlgorithm();
        return mortalityResult;
    }

    public DataTable Neighbors
    {
        get { return dtNeighbors; }
        set { dtNeighbors = value; }
    }
}
```

/// <summary>
/// 基株类；策略模式；
/// </summary>
public class Focal : Tree
{
    public Focal(Model model, Feature feature)
        : base(model, feature) { }
    public override Dictionary<string, double> GrowthAlgorithm()
    {
        Dictionary<string, double> growthDictionary = new Dictionary<string, double>();
        SiteParameter siteParameters = model.SiteParameters;
        double dGrowthHeight2DBH;
        double dI0;
        double dCompetitionIndex;
        double dAridityIndex;
        double dDEGDIndex;
        ModelBasicParameters(siteParameters, out dGrowthHeight2DBH, out dI0, out dCompetitionIndex, out dAridityIndex, out dDEGDIndex);
        double dPARResponseIntegral = model.PARResponseIntegral(dTreeHeight, dBoleHeight, dLeafArea, siteParameters.DeltaZ, dHalfSaturationPoint, dLightCompensationPoint, dI0, dCanopyArea, sPhenology);
        double dDBHGrowth = model.GrowthDBH(dCompetitionIndex, dDEGDIndex, dAridityIndex, dTreeHeight, dBoleHeight, dDBH, dGrowthHeight2DBH, dLeafArea, dSapwoodMaintenanceCostFactor, dGrowthScalingFactor, dPARResponseIntegral);
        double dLAGrowth = model.GrowthLeafArea(dDBHGrowth, dDBH, dLeafArea, dSapwoodTurnoverRate, dInitialLADDBH2Ratio);
        double dTreeHeightGrowth = model.GrowthTreeHeight(dHmax, dTreeHeight, dDBHGrowth, dDBH, dInitialSlopeDBHHeight);
        double dBoleHeightGrowth = model.GrowthBoleHeight(siteParameters.K, dBoleHeight, dI0, dLightCompensationPoint, dHalfSaturationPoint, dTreeHeight, dLeafArea, dCanopyArea, siteParameters.DeltaZ, sPhenology);
        growthDictionary.Add("TreeHeight", dTreeHeightGrowth);
        growthDictionary.Add("BoleHeight", dBoleHeightGrowth);

```
            growthDictionary.Add("DBH", dDBHGrowth);
            growthDictionary.Add("LeafArea", dLAGrowth);
            return growthDictionary;
        }


        /// <summary>
        /// 机理模型基本参数;
        /// </summary>
        /// <param name="siteParameters"></param>
        /// <param name="dGrowthHeight2DBH"></param>
        /// <param name="dI0"></param>
        /// <param name="dCompetitionIndex"></param>
        /// <param name="dAridityIndex"></param>
        /// <param name="dDEGDIndex"></param>
        private void ModelBasicParameters(SiteParameter siteParameters, out double dGrowthHeight2DBH, out double dI0, out double dCompetitionIndex, out double dAridityIndex, out double dDEGDIndex)
        {
            dGrowthHeight2DBH = model.GrowthHeight2DBH(dDBH, dHmax, dInitialSlopeDBHHeight);

            double dSLA = model.NeighborsSumLA(dtNeighbors, sSpecies, iiTopographicPosition, dTreeHeight, siteParameters.Phylodist);

            double dRadius = 0;
            if (siteParameters.NeighborhoodRadius == true)
            {
                dRadius = model.NeighborRadius(sSpecies, iiTopographicPosition, siteParameters.Phylodist);
            }
            else
            {
                dRadius = 7;
            }
```

```csharp
            double dSubquadratHillshade = model.HillshadeValueForIndividual
(dTreeX, dTreeY);
            dI0 = model.PARTreeTop(dSLA, dSubquadratHillshade, dRadius);
            double dTotalBA = model.TotalBA(dtNeighbors, sSpecies,
iiTopographicPosition, siteParameters.Phylodist);

            dTotalBA = dTotalBA + dBA;

            double dZOI = model.ZoneOfInfluence(dRadius);
            dCompetitionIndex = model.CompetitionIndex(dTotalBA, dZOI);
            dAridityIndex = model.AridityResponse(siteParameters.Aridity, dAridityMax);
            dDEGDIndex = model.DEGDResponse(dDEGDMax, dDEGDMin,
siteParameters.DEGD);
        }
        public override double MortalityAlgorithm()
        {
            SiteParameter siteParameters = model.SiteParameters;
            double dMortalityProbability = 0;
            double dGrowthHeight2DBH;
            double dI0;
            double dCompetitionIndex;
            double dAridityIndex;
            double dDEGDIndex;
            ModelBasicParameters(siteParameters, out dGrowthHeight2DBH, out
dI0, out dCompetitionIndex, out dAridityIndex, out dDEGDIndex);
            double dPARResponseIntegral = model.PARResponseIntegral
(dTreeHeight, dBoleHeight, dLeafArea, siteParameters.DeltaZ, dHalfSaturationPoint,
dLightCompensationPoint, dI0, dCanopyArea, sPhenology);
            double dGrowthEfficiency = model.GrowthEfficiency(dCompetitionIndex,
dDEGDIndex, dAridityIndex, dTreeHeight, dBoleHeight, dSapwoodMaintenanceCostFactor,
dGrowthScalingFactor, dLeafArea, siteParameters.DeltaZ, dHalfSaturationPoint,
dLightCompensationPoint, dI0, dCanopyArea, sPhenology);
            dIntrinsicMortalityRate = model.IntrinsicMortalityRate(dDBHmax, dDBH,
dIntrinsicMortalityRate);
            double dSoilMoisture = model.SoilMoistureValueForIndividual(dTreeX,
```

```csharp
dTreeY);
            dWindthrowMortalityRate = model.WindthrowMortalityRate(strRootType,
dSoilMoisture);
            dRandomMortalityRate = model.RandomMortalityRate();
            double dMortalityFunction = model.MortalityFunction(dIntrinsicMortalityRate,
dMortalityRateSuppression, dWindthrowMortalityRate, dRandomMortalityRate,
dGrowthEfficiency, dThresholdVigor);
            dMortalityProbability = model.MortalityProbability(dMortalityFunction,
siteParameters.TimeStep);
            return dMortalityProbability;
        }

    }


    /// <summary>
    /// 邻体类；策略模式;
    /// </summary>
    public class Neighbor : Tree
    {
        public Neighbor(Model model, Feature feature)
            : base(model, feature)
        { }
        public override Dictionary<string, double> GrowthAlgorithm()
        {
            MessageBox.Show("邻体不进行生长计算！");
            return null;
        }
        public override double MortalityAlgorithm()
        {
            MessageBox.Show("邻体不进行死亡计算！");
            return -9999;
        }

    }
```

附录 树木死亡动态模拟(TMDS)模型程序核心源代码(C#)

```csharp
/// <summary>
/// 树木死亡模拟批处理程序;
/// ODD protocol: Flow chart;
/// </summary>
public class TreeMortalityBatchProcess
{
    private Model model;
    private Layer oLayerFocals, oLayerNeighbor;
    private FeatureDefn oDefnFocals, oDefnNeighb;
    private Feature oFeatureBase, oFeatureNeighbor;
    private DataTable dtNeighbors, dtPhydist;
    private DataColumn dataColumn;
    private ToolStripProgressBar toolStripProgressBar1;
    private ToolStripStatusLabel toolStripStatusLabel1, toolStripStatusLabel2;

    public TreeMortalityBatchProcess(Model model, ToolStripProgressBar toolStripProgressBar1, ToolStripStatusLabel toolStripStatusLabel1, ToolStripStatusLabel toolStripStatusLabel2)
    {
        this.model = model;
        this.toolStripProgressBar1 = toolStripProgressBar1;
        this.toolStripStatusLabel1 = toolStripStatusLabel1;
        this.toolStripStatusLabel2 = toolStripStatusLabel2;

    }

    /// <summary>
    /// 初始化输入数据;
    /// </summary>
    /// <param name="strFocalsShp"></param>
    /// <param name="strNeighbShp"></param>
    private void InitializeGeospatialData(string strFocalsShp, string strNeighbShp, string strPhyloMatrix)
    {
        Gdal.SetConfigOption("SHAPE_ENCODING", "");
```

```
            Gdal.SetConfigOption ("GDAL_FILENAME_IS_UTF8", "YES");
            Ogr.RegisterAll ();
            Gdal.AllRegister ();
            DataSource dsFocals = Ogr.Open (strFocalsShp, 1);
            DataSource dsNeighbors = Ogr.Open (strNeighbShp, 0);
            oLayerFocals = dsFocals.GetLayerByIndex (0);
            oLayerFocals.ResetReading ();
            oDefnFocals = oLayerFocals.GetLayerDefn ();
            oLayerNeighbor = dsNeighbors.GetLayerByIndex (0);
            oLayerNeighbor.ResetReading ();
            oDefnNeighb = oLayerNeighbor.GetLayerDefn ();
            DataSet dsPhydist = Utility.PhylogeneticTreeFileToDataSet (strPhyloMatrix);
            dtPhydist = dsPhydist.Tables[0];
            dtPhydist.Columns["NoName"].ColumnName = "Species";
            dtPhydist.PrimaryKey = new DataColumn[] { dtPhydist.Columns["Species"]};
            toolStripProgressBar1.Maximum = Convert.ToInt32
(oLayerFocals.GetFeatureCount (1));
            toolStripProgressBar1.Value = 0;
            toolStripProgressBar1.Step = 1;
            toolStripStatusLabel1.Visible = true;
            toolStripStatusLabel2.Visible = true;
        }

        /// <summary>
        /// 批处理;
        /// ODD protocol: Flow chart;
        /// </summary>
        /// <param name="strFocalsShp"></param>
        /// <param name="strNeighbShp"></param>
        /// <param name="strPhyloMatrix"></param>
        public void BatchProcessing (string strFocalsShp, string strNeighbShp, string strPhyloMatrix, string strOutput)
        {
            for (int i = 0; i < model.SiteParameters.Years / model.SiteParameters.TimeStep; i++)
```

```csharp
        {
            InitializeGeospatialData(strFocalsShp, strNeighbShp, strPhyloMatrix);
            toolStripStatusLabel1.Text = "第" + Convert.ToInt32(i + 1) * model.SiteParameters.TimeStep + "年" + "/" + Convert.ToString(model.SiteParameters.Years / model.SiteParameters.TimeStep) + "年";
            System.Windows.Forms.Application.DoEvents();
            oLayerFocals.ResetReading();
            oLayerNeighbor.ResetReading();
            toolStripProgressBar1.Value = 0;
            ConstructNeighborsColumnsSchema();
            dtNeighbors.Rows.Clear();
            while ((oFeatureBase = oLayerFocals.GetNextFeature()) != null)
            {
                int fid = Convert.ToInt32(oFeatureBase.GetFID());
                if (fid == 481)
                {
                    int iddd = 0;
                }
                Tree focal = new Focal(model, oFeatureBase);
                Geometry geometryBuffer = oFeatureBase.GetGeometryRef().Buffer(10, 30);
                oLayerNeighbor.SetSpatialFilter(geometryBuffer);
                Geometry oGeometryBase = oFeatureBase.GetGeometryRef();
                dtNeighbors.Rows.Clear();
                FeatureEx featureBaseEx = new FeatureEx(oFeatureBase);

                long tick = DateTime.Now.Ticks;
                Random random = new Random((int)(tick & 0xffffffffL) | (int)(tick >> 32));
                int iRandomNumber = random.Next(50, 100);
                double dRandomMortalityRate = iRandomNumber * 0.01;
                if(Convert.ToDouble(featureBaseEx.GetFieldValue("SurvProb")) <dRandomMortalityRate)
                {
                    continue;
```

```
                        }
                    else
                    {
                        if (fid < 0)
                        {
                            toolStripProgressBar1.Value += toolStripProgressBar1.Step;
                            double dPercent = Convert.ToDouble(fid + 1) / Convert.ToDouble(oLayerFocals.GetFeatureCount(1));
                            dPercent = Math.Round(dPercent * 100, 1);
                            toolStripStatusLabel2.Text = Convert.ToString(dPercent) + "%";
                            System.Windows.Forms.Application.DoEvents();
                        }
                        else
                        {
                            toolStripProgressBar1.Value += toolStripProgressBar1.Step;
                            double dPercent = Convert.ToDouble(fid + 1) / Convert.ToDouble(oLayerFocals.GetFeatureCount(1));
                            dPercent = Math.Round(dPercent * 100, 1);
                            toolStripStatusLabel2.Text = Convert.ToString(dPercent) + "%";
                            System.Windows.Forms.Application.DoEvents();

                            while ((oFeatureNeighbor = oLayerNeighbor.GetNextFeature()) != null)
                            {
                                int fid2 = Convert.ToInt32(oFeatureNeighbor.GetFID());

                                FeatureEx featureNeighborEx = new FeatureEx(oFeatureNeighbor);

                                long tick2 = DateTime.Now.Ticks;
```

```csharp
                                Random random2 = new Random((int)(tick2
& 0xffffffffL) | (int)(tick2 >> 32));
                                int iRandomNumber2 = random2.Next(50, 100);
                                double dRandomMortalityRate2 =
iRandomNumber2 * 0.01;
                                double dRandomSurvProb = 1 -
dRandomMortalityRate2;
                                if
(Convert.ToDouble(featureNeighborEx.GetFieldValue("SurvProb")) > dRandomSurvProb)
                                {
                                    continue;
                                }
                                else
                                {
                                    if (fid != fid2)
                                    {
                                        double dSpatialDist =
oGeometryBase.Distance(oFeatureNeighbor.GetGeometryRef());
                                        string strBaseSpecies =
Convert.ToString(featureBaseEx.GetFieldValue("Species"));
                                        int iPosition =
Convert.ToInt32(featureBaseEx.GetFieldValue("Position"));
                                        double iNeighborRadius = 0;
                                        if
(model.SiteParameters.NeighborhoodRadius == true)
                                        {
                                            iNeighborRadius =
model.NeighborRadius(strBaseSpecies, iPosition, dtPhydist);
                                        }
                                        else
                                        {
                                            iNeighborRadius = 7;
                                        }

                                        if (dSpatialDist <= iNeighborRadius)
```

```
                                    {
                                        Tree neighbor = new Neighbor
(model, oFeatureNeighbor);
                                        DataRow dr =
dtPhydist.Rows.Find(strBaseSpecies);
                                        string strNeighborSpecies =
Convert.ToString(featureNeighborEx.GetFieldValue("Species"));
                                        double dPhyloDist =
Convert.ToDouble(dr[strNeighborSpecies].ToString());
                                        CreateNeigborRow
(featureNeighborEx, dSpatialDist, dPhyloDist);
                                    }
                                }
                            }
                        }

                        focal.Neighbors = dtNeighbors;

                        double dMortalityProbability = focal.Mortality();
                        if (i == 0)
                        {
                            double dSurvProb = 1 - dMortalityProbability;


                            long tick3 = DateTime.Now.Ticks;
                            Random random3 = new Random((int)(tick3
& 0xffffffffL) | (int)(tick3 >> 32));
                            int iRandomNumber3 = random3.Next(50, 100);
                            double dRandomMortalityRate3 =
iRandomNumber3 * 0.01;


                            double dPredMortProb = 1 - dSurvProb;
                            if (dPredMortProb > dRandomMortalityRate3)
                            {
```

```csharp
                        featureBaseEx.SetFieldValue("SurvProb", 0);
                    }
                    else
                    {
                        featureBaseEx.SetFieldValue("SurvProb", dSurvProb);
                    }
                }
                else
                {
                    double dPreviousSurvProb = Convert.ToDouble(featureBaseEx.GetFieldValue("SurvProb"));
                    double dSurvProb = dPreviousSurvProb * (1 - dMortalityProbability);



                    long tick4 = DateTime.Now.Ticks;
                    Random random4 = new Random((int)(tick4 & 0xffffffffL) | (int)(tick4 >> 32));
                    int iRandomNumber4 = random4.Next(50, 100);
                    double dRandomMortalityRate4 = iRandomNumber4 * 0.01;



                    double dPredMortProb = 1 - dSurvProb;
                    if (dPredMortProb > dRandomMortalityRate4)
                    {
                        featureBaseEx.SetFieldValue("SurvProb", 0);
                    }
                    else
                    {
                        featureBaseEx.SetFieldValue("SurvProb", dSurvProb);
                    }
```

```
                            }

                            Dictionary<string, double> growthDictionary =
focal.Growth();
                            double dTreeHeightGrowth =
growthDictionary["TreeHeight"];
                            double dBoleHeightGrowth =
growthDictionary["BoleHeight"];
                            double dDBHGrowth = growthDictionary["DBH"];
                            double dLeafAreaGrowth = growthDictionary
["LeafArea"];
                            featureBaseEx.SetFieldValue("TreeHeight",
dTreeHeightGrowth + Convert.ToDouble(featureBaseEx.GetFieldValue("TreeHeight")));
                            featureBaseEx.SetFieldValue("BoleHeight",
dBoleHeightGrowth + Convert.ToDouble(featureBaseEx.GetFieldValue("BoleHeight")));
                            featureBaseEx.SetFieldValue("DBH", dDBHGrowth
+ Convert.ToDouble(featureBaseEx.GetFieldValue("DBH")));
                            featureBaseEx.SetFieldValue("LeafArea",
dLeafAreaGrowth + Convert.ToDouble(featureBaseEx.GetFieldValue("LeafArea")));
                            oLayerFocals.SetFeature(oFeatureBase);
                            oLayerNeighbor.ResetReading();

                        }
                    }
                }


                string strFolderName = Path.GetDirectoryName(strNeighbShp);
                string strFileName = Path.GetFileNameWithoutExtension
(strNeighbShp);
                string strNeighborDbf = strFolderName + "\\" + strFileName + ".dbf";
                string strNeighborShx = strFolderName + "\\" + strFileName + ".shx";
                System.IO.File.Copy(strFocalsShp, strNeighbShp, true);
                System.IO.File.Copy(strFolderName + "\\" + "Focal.dbf",
strNeighborDbf, true);
```

```csharp
                System.IO.File.Copy(strFolderName + "\\" + "Focal.shx",
strNeighborShx, true);
            }

                string strOutputFolderName = Path.GetDirectoryName(strOutput);
                string strOutputFileName = Path.GetFileNameWithoutExtension(strOutput);
                string strOutputDbf = strOutputFolderName + "\\" + strOutputFileName +
".dbf";
                string strOutputShx = strOutputFolderName + "\\" + strOutputFileName +
".shx";
                string strFocalFolderName = Path.GetDirectoryName(strFocalsShp);
                string strFocalFileName = Path.GetFileNameWithoutExtension(strFocalsShp);
                System.IO.File.Copy(strFocalsShp, strOutput, true);
                System.IO.File.Copy(strFocalFolderName + "\\" + strFocalFileName +
".dbf", strOutputDbf, true);
                System.IO.File.Copy(strFocalFolderName + "\\" + strFocalFileName +
".shx", strOutputShx, true);
                MessageBox.Show("OK");
                toolStripProgressBar1.Visible = false;
                toolStripStatusLabel1.Visible = false;
                toolStripStatusLabel2.Visible = false;
            }




        /// <summary>
        /// 创建邻体结合一行记录;
        /// </summary>
        private void CreateNeigborRow(FeatureEx featureNeighborEx, double
dSpatialDist, double dPhyloDist)
        {
                string strTag = Convert.ToString(featureNeighborEx.GetFieldValue("tag"));
                string strSpecies =
Convert.ToString(featureNeighborEx.GetFieldValue("Species"));
                double dTreeHeight =
Convert.ToDouble(featureNeighborEx.GetFieldValue("TreeHeight"));
```

```csharp
            double dBA = Convert.ToDouble(featureNeighborEx.GetFieldValue("BA"));
            double dLeafArea =
Convert.ToDouble(featureNeighborEx.GetFieldValue("LeafArea"));
            double dSurvProbability =
Convert.ToDouble(featureNeighborEx.GetFieldValue("SurvProb"));
            DataRow dataRow = dtNeighbors.NewRow();
            dataRow["tag"] = strTag;
            dataRow["Species"] = strSpecies;
            dataRow["TreeHeight"] = dTreeHeight;
            dataRow["BA"] = dBA;
            dataRow["SpatialDistance"] = dSpatialDist;
            dataRow["PhyloDistance"] = dPhyloDist;
            dataRow["LeafArea"] = dLeafArea;
            dataRow["SurvProb"] = dSurvProbability;
            dtNeighbors.Rows.Add(dataRow);
    }


    /// <summary>
    /// 构建邻体集合列架构;
    /// </summary>
    private void ConstructNeighborsColumnsSchema()
    {
        dtNeighbors = new DataTable("Neighbors");
        dataColumn = new DataColumn("tag", Type.GetType("System.String"));
        dtNeighbors.Columns.Add(dataColumn);
        dataColumn = new DataColumn("Species", Type.GetType("System.String"));
        dtNeighbors.Columns.Add(dataColumn);
        dataColumn = new DataColumn("TreeHeight",
Type.GetType("System.Double"));
        dtNeighbors.Columns.Add(dataColumn);
        dataColumn = new DataColumn("BA", Type.GetType("System.Double"));
        dtNeighbors.Columns.Add(dataColumn);
        dataColumn = new DataColumn("SpatialDistance",
Type.GetType("System.Double"));
```

```csharp
            dtNeighbors.Columns.Add(dataColumn);
            dataColumn = new DataColumn("PhyloDistance", Type.GetType("System.Double"));
            dtNeighbors.Columns.Add(dataColumn);
            dataColumn = new DataColumn("LeafArea", Type.GetType("System.Double"));
            dtNeighbors.Columns.Add(dataColumn);
            dataColumn = new DataColumn("SurvProb", Type.GetType("System.Double"));
            dtNeighbors.Columns.Add(dataColumn);
        }
    }



    /// <summary>
    /// 封装 Feature 类，便于读取字段值;
    /// </summary>
    public class FeatureEx
    {
        private Feature feature;
        public FeatureEx(Feature feature)
        {
            this.feature = feature;
        }

        public object GetFieldValue(string strFieldName)
        {
            object oFieldValue = null;
            int iFieldIndex = feature.GetDefnRef().GetFieldIndex(strFieldName);
            FieldType fieldType = feature.GetFieldType(strFieldName);
            if (fieldType == FieldType.OFTInteger)
            {
                oFieldValue = feature.GetFieldAsInteger(iFieldIndex);
            }
            else if (fieldType == FieldType.OFTReal)
```

```csharp
        {
            oFieldValue = feature.GetFieldAsDouble(iFieldIndex);
        }
        else if (fieldType == FieldType.OFTString)
        {
            oFieldValue = feature.GetFieldAsString(iFieldIndex);
        }
        else
        {
            return null;
        }
        return oFieldValue;
    }

    public void SetFieldValue(string strFieldName, object oFieldValue)
    {
        int iFieldIndex = feature.GetDefnRef().GetFieldIndex(strFieldName);
        FieldType fieldType = feature.GetFieldType(strFieldName);
        if (fieldType == FieldType.OFTInteger)
        {
            feature.SetField(iFieldIndex, Convert.ToInt32(oFieldValue));
        }
        else if (fieldType == FieldType.OFTReal)
        {
            feature.SetField(iFieldIndex, Convert.ToDouble(oFieldValue));
        }
        else if (fieldType == FieldType.OFTString)
        {
            feature.SetField(iFieldIndex, Convert.ToString(oFieldValue));
        }
        else
        {
            // null;
        }
```

```csharp
        }

    }



    /// <summary>
    /// 创建 Shapefile 接口;
    /// </summary>
    public interface ICreateEsriShapefile
    {
        bool CreateEsriShapefileFromExcel(string strExcel, string strShape);
    }


    /// <summary>
    /// 创建树木点要素类;
    /// </summary>
    class CreateTreePointFeatureClass : ICreateEsriShapefile
    {
        public bool CreateEsriShapefileFromExcel(string strExcel, string strShape)
        {
            try
            {
                Gdal.SetConfigOption("GDAL_FILENAME_IS_UTF8", "YES");
                Gdal.SetConfigOption("SHAPE_ENCODING", "");
                string strFile = strShape;
                Ogr.RegisterAll();
                string strDriverName = "ESRI Shapefile";
                OSGeo.OGR.Driver oDriver = Ogr.GetDriverByName(strDriverName);
                DataSource oDS = oDriver.CreateDataSource(strFile, null);
                Layer oLayer = oDS.CreateLayer("TreePoint", null, wkbGeometryType.wkbPoint, null);
                CreatePointFields(oLayer);
                FeatureDefn oDefn = oLayer.GetLayerDefn();
                System.Data.DataSet dsTree = Utility.ExcelToDataSet(strExcel, "Sheet1");
```

```
System.Data.DataTable dtTree = dsTree.Tables[0];
for (int i = 0; i < dtTree.Rows.Count; i++)
{
    System.Windows.Forms.Application.DoEvents();
    string strTag = Convert.ToString(dtTree.Rows[i]["tag"].ToString());
    string strSpecies = Convert.ToString(dtTree.Rows[i]["Species"].ToString());
    double dDBH = Convert.ToDouble(dtTree.Rows[i]["DBH"].ToString());
    double dBA = Convert.ToDouble(dtTree.Rows[i]["BA"].ToString());
    double dCW = Convert.ToDouble(dtTree.Rows[i]["CW"].ToString());
    double dTreeHeight = Convert.ToDouble(dtTree.Rows[i]["TreeHeight"].ToString());
    double dBoleHeight = Convert.ToDouble(dtTree.Rows[i]["BoleHeight"].ToString());
    double dCrownHeight = Convert.ToDouble(dtTree.Rows[i]["CrownHeight"].ToString());
    double dgx = Convert.ToDouble(dtTree.Rows[i]["gx"].ToString());
    double dgy = Convert.ToDouble(dtTree.Rows[i]["gy"].ToString());
    double dSurvival = Convert.ToDouble(dtTree.Rows[i]["Survival"].ToString());
    double dPosition = Convert.ToDouble(dtTree.Rows[i]["Position"].ToString());
    double dElevation = Convert.ToDouble(dtTree.Rows[i]["Elevation"].ToString());
    //double dHillshade = Convert.ToDouble(dtTree.Rows[i]["Hillshade"].ToString());
    double dConvexity = Convert.ToDouble(dtTree.Rows[i]["Convexity"].ToString());
    double dLAD2Ratio =
```

```csharp
Convert.ToDouble(dtTree.Rows[i]["LAD2Ratio"].ToString());
                double dHmax =
Convert.ToDouble(dtTree.Rows[i]["Hmax"].ToString());
                //double dDBHmax =
Convert.ToDouble(dtTree.Rows[i]["DBHmax"].ToString());
                double dSlopeDH =
Convert.ToDouble(dtTree.Rows[i]["SlopeDH"].ToString());
                double dCompensation =
Convert.ToDouble(dtTree.Rows[i]["Compensation"].ToString());
                double dSaturation =
Convert.ToDouble(dtTree.Rows[i]["Saturation"].ToString());
                double dGrowScale =
Convert.ToDouble(dtTree.Rows[i]["GrowScale"].ToString());
                double dSapwMaint =
Convert.ToDouble(dtTree.Rows[i]["SapwMaint"].ToString());
                double dSapwTurn =
Convert.ToDouble(dtTree.Rows[i]["SapwTurn"].ToString());
                double dIntrinsicM =
Convert.ToDouble(dtTree.Rows[i]["IntrinsicM"].ToString());
                double dSuppression =
Convert.ToDouble(dtTree.Rows[i]["Suppression"].ToString());
                double dDEGDMin =
Convert.ToDouble(dtTree.Rows[i]["DEGDMin"].ToString());
                double dDEGDMax =
Convert.ToDouble(dtTree.Rows[i]["DEGDMax"].ToString());
                double dAridityMax =
Convert.ToDouble(dtTree.Rows[i]["AridityMax"].ToString());
                double dSLA =
Convert.ToDouble(dtTree.Rows[i]["SLA"].ToString());
                double dBiomassA =
Convert.ToDouble(dtTree.Rows[i]["BiomassA"].ToString());
                double dBiomassB =
Convert.ToDouble(dtTree.Rows[i]["BiomassB"].ToString());
                double dLeafBiomass =
Convert.ToDouble(dtTree.Rows[i]["LeafBiomass"].ToString());
```

```
                    double dLeafArea =
Convert.ToDouble(dtTree.Rows[i]["LeafArea"].ToString());
                    double dThresholdV =
Convert.ToDouble(dtTree.Rows[i]["ThresholdV"].ToString());
                    double dMortProb =
Convert.ToDouble(dtTree.Rows[i]["SurvProb"].ToString());

                    Feature oFeaturePoint = new Feature(oDefn);
                    oFeaturePoint.SetField(0, strTag);
                    oFeaturePoint.SetField(1, strSpecies);
                    oFeaturePoint.SetField(2, dDBH);
                    oFeaturePoint.SetField(3, dBA);
                    oFeaturePoint.SetField(4, dCW);
                    oFeaturePoint.SetField(5, dTreeHeight);
                    oFeaturePoint.SetField(6, dBoleHeight);
                    oFeaturePoint.SetField(7, dCrownHeight);
                    oFeaturePoint.SetField(8, dgx);
                    oFeaturePoint.SetField(9, dgy);
                    oFeaturePoint.SetField(10, dSurvival);
                    oFeaturePoint.SetField(11, dPosition);
                    oFeaturePoint.SetField(12, dElevation);
                    //oFeaturePoint.SetField(13, dHillshade);
                    oFeaturePoint.SetField(14, dConvexity);
                    oFeaturePoint.SetField(15, dLAD2Ratio);
                    oFeaturePoint.SetField(16, dHmax);
                    oFeaturePoint.SetField(17, dSlopeDH);
                    oFeaturePoint.SetField(18, dCompensation);
                    oFeaturePoint.SetField(19, dSaturation);
                    oFeaturePoint.SetField(20, dGrowScale);
                    oFeaturePoint.SetField(21, dSapwMaint);
                    oFeaturePoint.SetField(22, dSapwTurn);
                    oFeaturePoint.SetField(23, dIntrinsicM);
                    oFeaturePoint.SetField(24, dSuppression);
                    oFeaturePoint.SetField(25, dDEGDMin);
                    oFeaturePoint.SetField(26, dDEGDMax);
```

```csharp
                oFeaturePoint.SetField(27, dAridityMax);
                oFeaturePoint.SetField(28, dSLA);
                oFeaturePoint.SetField(29, dBiomassA);
                oFeaturePoint.SetField(30, dBiomassB);
                oFeaturePoint.SetField(31, dLeafBiomass);
                oFeaturePoint.SetField(32, dLeafArea);
                oFeaturePoint.SetField(33, dThresholdV);
                oFeaturePoint.SetField(34, dMortProb);
                Geometry geomPoint = Geometry.CreateFromWkt("POINT (" + dgx + " " + dgy + ")");
                oFeaturePoint.SetGeometry(geomPoint);
                oLayer.CreateFeature(oFeaturePoint);
            }
        }
        catch (Exception ex)
        {
            MessageBox.Show(ex.ToString() + "也有可能是OGR的BUG所致！");
            return false;
        }
        return true;
    }


    /// <summary>
    /// 创建树木字段；
    /// </summary>
    /// <param name="oLayer"></param>
    private static void CreatePointFields(Layer oLayer)
    {
        FieldDefn oTagFieldDefn = new FieldDefn("tag", FieldType.OFTString);
        oTagFieldDefn.SetWidth(100);
        oTagFieldDefn.SetPrecision(6);
        oLayer.CreateField(oTagFieldDefn, 1);
        FieldDefn oSpeciesFieldDefn = new FieldDefn("Species",
```

FieldType.OFTString);
    oSpeciesFieldDefn.SetWidth(100);
    oSpeciesFieldDefn.SetPrecision(6);
    oLayer.CreateField(oSpeciesFieldDefn, 1);
    FieldDefn oDBHFieldDefn = new FieldDefn("DBH", FieldType.OFTReal);
    oDBHFieldDefn.SetWidth(100);
    oDBHFieldDefn.SetPrecision(6);
    oLayer.CreateField(oDBHFieldDefn, 1);
    FieldDefn oBAFieldDefn = new FieldDefn("BA", FieldType.OFTReal);
    oBAFieldDefn.SetWidth(100);
    oBAFieldDefn.SetPrecision(6);
    oLayer.CreateField(oBAFieldDefn, 1);
    FieldDefn oCWFieldDefn = new FieldDefn("CW", FieldType.OFTReal);
    oCWFieldDefn.SetWidth(100);
    oCWFieldDefn.SetPrecision(6);
    oLayer.CreateField(oCWFieldDefn, 1);
    FieldDefn oTreeHeightFieldDefn = new FieldDefn("TreeHeight", FieldType.OFTReal);
    oTreeHeightFieldDefn.SetWidth(100);
    oTreeHeightFieldDefn.SetPrecision(6);
    oLayer.CreateField(oTreeHeightFieldDefn, 1);
    FieldDefn oBoleHeightFieldDefn = new FieldDefn("BoleHeight", FieldType.OFTReal);
    oBoleHeightFieldDefn.SetWidth(100);
    oBoleHeightFieldDefn.SetPrecision(6);
    oLayer.CreateField(oBoleHeightFieldDefn, 1);
    FieldDefn oCrownHeightFieldDefn = new FieldDefn("CrownHeight", FieldType.OFTReal);
    oCrownHeightFieldDefn.SetWidth(100);
    oCrownHeightFieldDefn.SetPrecision(6);
    oLayer.CreateField(oCrownHeightFieldDefn, 1);
    FieldDefn oGxFieldDefn = new FieldDefn("gx", FieldType.OFTReal);
    oGxFieldDefn.SetWidth(100);
    oGxFieldDefn.SetPrecision(6);
    oLayer.CreateField(oGxFieldDefn, 1);

```csharp
FieldDefn oGyFieldDefn = new FieldDefn ("gy", FieldType.OFTReal);
oGyFieldDefn.SetWidth (100);
oGyFieldDefn.SetPrecision (6);
oLayer.CreateField (oGyFieldDefn, 1);
FieldDefn oSurvivalFieldDefn = new FieldDefn ("Survival", FieldType.OFTReal);
oSurvivalFieldDefn.SetWidth (100);
oSurvivalFieldDefn.SetPrecision (6);
oLayer.CreateField (oSurvivalFieldDefn, 1);
FieldDefn oPositionFieldDefn = new FieldDefn ("Position", FieldType.OFTReal);
oPositionFieldDefn.SetWidth (100);
oPositionFieldDefn.SetPrecision (6);
oLayer.CreateField (oPositionFieldDefn, 1);
FieldDefn oElevationFieldDefn = new FieldDefn ("Elevation", FieldType.OFTReal);
oElevationFieldDefn.SetWidth (100);
oElevationFieldDefn.SetPrecision (6);
oLayer.CreateField (oElevationFieldDefn, 1);
//FieldDefn oHillshadeFieldDefn = new FieldDefn ("Hillshade", FieldType.OFTReal);
//oHillshadeFieldDefn.SetWidth (100);
//oHillshadeFieldDefn.SetPrecision (6);
//oLayer.CreateField (oHillshadeFieldDefn, 1);
FieldDefn oConvexityFieldDefn = new FieldDefn ("Convexity", FieldType.OFTReal);
oConvexityFieldDefn.SetWidth (100);
oConvexityFieldDefn.SetPrecision (6);
oLayer.CreateField (oConvexityFieldDefn, 1);
FieldDefn oLAD2RatioFieldDefn = new FieldDefn ("LAD2Ratio", FieldType.OFTReal);
oLAD2RatioFieldDefn.SetWidth (100);
oLAD2RatioFieldDefn.SetPrecision (6);
oLayer.CreateField (oLAD2RatioFieldDefn, 1);
FieldDefn oHmaxFieldDefn = new FieldDefn ("Hmax", FieldType.OFTReal);
```

```
oHmaxFieldDefn.SetWidth(100);
oHmaxFieldDefn.SetPrecision(6);
oLayer.CreateField(oHmaxFieldDefn, 1);
FieldDefn oSlopeDHFieldDefn = new FieldDefn("SlopeDH", FieldType.OFTReal);
oSlopeDHFieldDefn.SetWidth(100);
oSlopeDHFieldDefn.SetPrecision(6);
oLayer.CreateField(oSlopeDHFieldDefn, 1);
FieldDefn oCompensationFieldDefn = new FieldDefn("Compensation", FieldType.OFTReal);
oCompensationFieldDefn.SetWidth(100);
oCompensationFieldDefn.SetPrecision(6);
oLayer.CreateField(oCompensationFieldDefn, 1);
FieldDefn oSaturationFieldDefn = new FieldDefn("Saturation", FieldType.OFTReal);
oSaturationFieldDefn.SetWidth(100);
oSaturationFieldDefn.SetPrecision(6);
oLayer.CreateField(oSaturationFieldDefn, 1);
FieldDefn oGrowScaleFieldDefn = new FieldDefn("GrowScale", FieldType.OFTReal);
oGrowScaleFieldDefn.SetWidth(100);
oGrowScaleFieldDefn.SetPrecision(6);
oLayer.CreateField(oGrowScaleFieldDefn, 1);
FieldDefn oSapwMaintFieldDefn = new FieldDefn("SapwMaint", FieldType.OFTReal);
oSapwMaintFieldDefn.SetWidth(100);
oSapwMaintFieldDefn.SetPrecision(6);
oLayer.CreateField(oSapwMaintFieldDefn, 1);
FieldDefn oSapwTurnFieldDefn = new FieldDefn("SapwTurn", FieldType.OFTReal);
oSapwTurnFieldDefn.SetWidth(100);
oSapwTurnFieldDefn.SetPrecision(6);
oLayer.CreateField(oSapwTurnFieldDefn, 1);
FieldDefn oIntrinsicMFieldDefn = new FieldDefn("IntrinsicM", FieldType.OFTReal);
```

```csharp
                oIntrinsicMFieldDefn.SetWidth(100);
                oIntrinsicMFieldDefn.SetPrecision(6);
                oLayer.CreateField(oIntrinsicMFieldDefn, 1);
                FieldDefn oSuppressionFieldDefn = new FieldDefn("Suppression",
FieldType.OFTReal);
                oSuppressionFieldDefn.SetWidth(100);
                oSuppressionFieldDefn.SetPrecision(6);
                oLayer.CreateField(oSuppressionFieldDefn, 1);
                FieldDefn oDEGDMinFieldDefn = new FieldDefn("DEGDMin",
FieldType.OFTReal);
                oDEGDMinFieldDefn.SetWidth(100);
                oDEGDMinFieldDefn.SetPrecision(6);
                oLayer.CreateField(oDEGDMinFieldDefn, 1);
                FieldDefn oDEGDMaxFieldDefn = new FieldDefn("DEGDMax",
FieldType.OFTReal);
                oDEGDMaxFieldDefn.SetWidth(100);
                oDEGDMaxFieldDefn.SetPrecision(6);
                oLayer.CreateField(oDEGDMaxFieldDefn, 1);
                FieldDefn oAridityMaxFieldDefn = new FieldDefn("AridityMax",
FieldType.OFTReal);
                oAridityMaxFieldDefn.SetWidth(100);
                oAridityMaxFieldDefn.SetPrecision(6);
                oLayer.CreateField(oAridityMaxFieldDefn, 1);
                FieldDefn oSLAFieldDefn = new FieldDefn("SLA", FieldType.OFTReal);
                oSLAFieldDefn.SetWidth(100);
                oSLAFieldDefn.SetPrecision(6);
                oLayer.CreateField(oSLAFieldDefn, 1);
                FieldDefn oBiomassAFieldDefn = new FieldDefn("BiomassA",
FieldType.OFTReal);
                oBiomassAFieldDefn.SetWidth(100);
                oBiomassAFieldDefn.SetPrecision(6);
                oLayer.CreateField(oBiomassAFieldDefn, 1);
                FieldDefn oBiomassBFieldDefn = new FieldDefn("BiomassB",
FieldType.OFTReal);
                oBiomassBFieldDefn.SetWidth(100);
```

oBiomassBFieldDefn.SetPrecision(6);
oLayer.CreateField(oBiomassBFieldDefn, 1);
FieldDefn oLeafBiomassFieldDefn = new FieldDefn("LeafBiomass", FieldType.OFTReal);
oLeafBiomassFieldDefn.SetWidth(100);
oLeafBiomassFieldDefn.SetPrecision(6);
oLayer.CreateField(oLeafBiomassFieldDefn, 1);
FieldDefn oLeafAreaFieldDefn = new FieldDefn("LeafArea", FieldType.OFTReal);
oLeafAreaFieldDefn.SetWidth(100);
oLeafAreaFieldDefn.SetPrecision(6);
oLayer.CreateField(oLeafAreaFieldDefn, 1);
FieldDefn oThresholdVFieldDefn = new FieldDefn("ThresholdV", FieldType.OFTReal);
oThresholdVFieldDefn.SetWidth(100);
oThresholdVFieldDefn.SetPrecision(6);
oLayer.CreateField(oThresholdVFieldDefn, 1);
FieldDefn oMortProbFieldDefn = new FieldDefn("SurvProb", FieldType.OFTReal);
oMortProbFieldDefn.SetWidth(100);
oMortProbFieldDefn.SetPrecision(6);
oLayer.CreateField(oMortProbFieldDefn, 1);

}

}

/// <summary>
/// 创建树冠投影多边形要素类；
/// </summary>
class CreateCanopyPolygonFeatureClass : ICreateEsriShapefile
{
    public bool CreateEsriShapefileFromExcel(string strExcel, string strShape)
    {
        try

```csharp
{
    Gdal.SetConfigOption ("GDAL_FILENAME_IS_UTF8", "YES");
    Gdal.SetConfigOption ("SHAPE_ENCODING", "");
    string strFile = strShape;
    Ogr.RegisterAll ();
    string strDriverName = "ESRI Shapefile";
    OSGeo.OGR.Driver oDriver = Ogr.GetDriverByName (strDriverName);
    DataSource oDS = oDriver.CreateDataSource (strFile, null);
    Layer oLayer = oDS.CreateLayer ("CanopyPolygon", null, wkbGeometryType.wkbMultiPolygon, null);
    CreatePolygonFields (oLayer);
    FeatureDefn oDefn = oLayer.GetLayerDefn ();
    System.Data.DataSet dsCanopy = Utility.ExcelToDataSet (strExcel, "Sheet1");
    System.Data.DataTable dtCanopy = dsCanopy.Tables[0];
    for (int i = 0; i < dtCanopy.Rows.Count; i++)
    {
        System.Windows.Forms.Application.DoEvents ();
        string strTag = Convert.ToString (dtCanopy.Rows[i]["tag"].ToString ());
        double dN = Convert.ToDouble (dtCanopy.Rows[i]["N"].ToString ());
        double dNE = Convert.ToDouble (dtCanopy.Rows[i]["NE"].ToString ());
        double dNW = Convert.ToDouble (dtCanopy.Rows[i]["NW"].ToString ());
        double dW = Convert.ToDouble (dtCanopy.Rows[i]["W"].ToString ());
        double dSW = Convert.ToDouble (dtCanopy.Rows[i]["SW"].ToString ());
        double dS = Convert.ToDouble (dtCanopy.Rows[i]["S"].ToString ());
        double dSE = Convert.ToDouble (dtCanopy.Rows[i]["SE"].ToString ());
        double dE =
```

Convert.ToDouble(dtCanopy.Rows[i]["E"].ToString());
  double dGX =
Convert.ToDouble(dtCanopy.Rows[i]["gx"].ToString());
  double dGY =
Convert.ToDouble(dtCanopy.Rows[i]["gy"].ToString());

  Feature oFeaturePolygon = new Feature(oDefn);
  oFeaturePolygon.SetField(0, strTag);
  oFeaturePolygon.SetField(1, dN);
  oFeaturePolygon.SetField(2, dNE);
  oFeaturePolygon.SetField(3, dNW);
  oFeaturePolygon.SetField(4, dW);
  oFeaturePolygon.SetField(5, dSW);
  oFeaturePolygon.SetField(6, dS);
  oFeaturePolygon.SetField(7, dSE);
  oFeaturePolygon.SetField(8, dE);
  oFeaturePolygon.SetField(9, dGX);
  oFeaturePolygon.SetField(10, dGY);

  double dNgx = 0, dNgy = 0, dNEgx = 0, dNEgy = 0, dNWgx,
dNWgy, dWgx, dWgy, dSWgx, dSWgy, dSgx, dSgy, dSEgx, dSEgy, dEgx, dEgy = 0;
  int iAngle = 45;
  double dCanopyArea = 0;
  dNgx = dGX;
  dNgy = dGY + dN;
  dNEgx = dGX + dNE * Math.Sin(iAngle);
  dNEgy = dGY + dNE * Math.Cos(iAngle);
  dNWgx = dGX – dNW * Math.Sin(iAngle);
  dNWgy = dGY + dNW * Math.Cos(iAngle);
  dWgx = dGX – dW;
  dWgy = dGY;
  dSWgx = dGX – dSW * Math.Sin(iAngle);
  dSWgy = dGY – dSW * Math.Cos(iAngle);
  dSgx = dGX;
  dSgy = dGY – dS;

```csharp
dSEgx = dGX + dSE * Math.Sin(iAngle);
dSEgy = dGY – dSE * Math.Cos(iAngle);
dEgx = dGX + dE;
dEgy = dGY;
Geometry arcAngle1 = null;
Geometry arcAngle2 = null;
Geometry arcAngle3 = null;
Geometry arcAngle4 = null;
if (dN > dE)
{
    arcAngle1 = Ogr.ApproximateArcAngles(dGX, dGY, 0, dN, dE, -90, 0, 90, 1);
}
else
{
    arcAngle1 = Ogr.ApproximateArcAngles(dGX, dGY, 0, dE, dN, 0, -90, 0, 1);
}
if (dE > dS)
{
    arcAngle2 = Ogr.ApproximateArcAngles(dGX, dGY, 0, dE, dS, 0, 90, 0, 1);
}
else
{
    arcAngle2 = Ogr.ApproximateArcAngles(dGX, dGY, 0, dS, dE, 90, -90, 0, 1);
}
if (dS > dW)
{
    arcAngle3 = Ogr.ApproximateArcAngles(dGX, dGY, 0, dS, dW, 90, 0, 90, 1);
}
else
```

```
            {
                arcAngle3 = Ogr.ApproximateArcAngles(dGX, dGY, 0, dW, dS, 0, 90, 180, 1);

            }
            if (dW > dN)
            {
                arcAngle4 = Ogr.ApproximateArcAngles(dGX, dGY, 0, dW, dN, 0, 180, 270, 1);
            }
            else
            {
                arcAngle4 = Ogr.ApproximateArcAngles(dGX, dGY, 0, dN, dW, -90, -90, 0, 1);
            }

            arcAngle3 = arcAngle3.Union(arcAngle4);
            arcAngle2 = arcAngle2.Union(arcAngle3);
            arcAngle1 = arcAngle1.Union(arcAngle2);
            arcAngle1 = arcAngle1.ConvexHull();
            Geometry geometryPolygon = arcAngle1;
            geometryPolygon.CloseRings();
            oFeaturePolygon.SetGeometry(geometryPolygon);
            dCanopyArea = geometryPolygon.GetArea();
            oFeaturePolygon.SetField(11, dCanopyArea);
            oLayer.CreateFeature(oFeaturePolygon);
        }
    }
    catch (Exception ex)
    {
        MessageBox.Show(ex.ToString() + "也有可能是OGR的BUG所致！");
        return false;
    }
    return true;
```

}

/// <summary>
/// 创建冠层字段;
/// </summary>
/// <param name="oLayer"></param>
private static void CreatePolygonFields(Layer oLayer)
{
    FieldDefn oTagFieldDefn = new FieldDefn("tag", FieldType.OFTString);
    oTagFieldDefn.SetWidth(100);
    oTagFieldDefn.SetPrecision(6);
    oLayer.CreateField(oTagFieldDefn, 1);
    FieldDefn oNFieldDefn = new FieldDefn("N", FieldType.OFTReal);
    oNFieldDefn.SetWidth(100);
    oNFieldDefn.SetPrecision(6);
    oLayer.CreateField(oNFieldDefn, 1);
    FieldDefn oNEFieldDefn = new FieldDefn("NE", FieldType.OFTReal);
    oNEFieldDefn.SetWidth(100);
    oNEFieldDefn.SetPrecision(6);
    oLayer.CreateField(oNEFieldDefn, 1);
    FieldDefn oNWFieldDefn = new FieldDefn("NW", FieldType.OFTReal);
    oNWFieldDefn.SetWidth(100);
    oNWFieldDefn.SetPrecision(6);
    oLayer.CreateField(oNWFieldDefn, 1);
    FieldDefn oWFieldDefn = new FieldDefn("W", FieldType.OFTReal);
    oWFieldDefn.SetWidth(100);
    oWFieldDefn.SetPrecision(6);
    oLayer.CreateField(oWFieldDefn, 1);
    FieldDefn oSWEFieldDefn = new FieldDefn("SW", FieldType.OFTReal);
    oSWEFieldDefn.SetWidth(100);
    oSWEFieldDefn.SetPrecision(6);
    oLayer.CreateField(oSWEFieldDefn, 1);
    FieldDefn oSFieldDefn = new FieldDefn("S", FieldType.OFTReal);
    oSFieldDefn.SetWidth(100);
    oSFieldDefn.SetPrecision(6);

```
            oLayer.CreateField(oSFieldDefn, 1);
            FieldDefn oSEFieldDefn = new FieldDefn("SE", FieldType.OFTReal);
            oSEFieldDefn.SetWidth(100);
            oSEFieldDefn.SetPrecision(6);
            oLayer.CreateField(oSEFieldDefn, 1);
            FieldDefn oEFieldDefn = new FieldDefn("E", FieldType.OFTReal);
            oEFieldDefn.SetWidth(100);
            oEFieldDefn.SetPrecision(6);
            oLayer.CreateField(oEFieldDefn, 1);
            FieldDefn oGXFieldDefn = new FieldDefn("gx", FieldType.OFTReal);
            oGXFieldDefn.SetWidth(100);
            oGXFieldDefn.SetPrecision(6);
            oLayer.CreateField(oGXFieldDefn, 1);
            FieldDefn oGYFieldDefn = new FieldDefn("gy", FieldType.OFTReal);
            oGYFieldDefn.SetWidth(100);
            oGYFieldDefn.SetPrecision(6);
            oLayer.CreateField(oGYFieldDefn, 1);
            FieldDefn oCanopyAreaFieldDefn = new FieldDefn("CanopyArea", FieldType.OFTReal);
            oCanopyAreaFieldDefn.SetWidth(100);
            oCanopyAreaFieldDefn.SetPrecision(6);
            oLayer.CreateField(oCanopyAreaFieldDefn, 1);

        }
    }


    /// <summary>
    /// 模型模拟结果类;
    /// </summary>
    public class SimulationResult
    {
        REngine rEngine = null;
        public SimulationResult()
        {
            REngine.SetEnvironmentVariables(@"D:\R\R-3.4.4\bin\x64",
```

```csharp
            @"D:\R\R-3.4.4");
            rEngine = REngine.GetInstance();
            rEngine.Initialize();
            rEngine.Evaluate("library(pROC)");
            rEngine.Evaluate("library(maptools)");
            rEngine.Evaluate("library(ggplot2)");
            rEngine.Evaluate("library(sp)");
            rEngine.Evaluate("library(RColorBrewer)");
        }

        /// <summary>
        /// REngineDispose;
        /// </summary>
        public void REngineDispose()
        {
            rEngine.Dispose();
        }

        /// <summary>
        /// 调用 pROC 程序包计算 AUC;
        /// </summary>
        /// <param name="dataTableInput"></param>
        /// <returns></returns>
        public double AreaUnderROCCurve(System.Data.DataTable dataTableInput)
        {
            double dAUC = 0;
            System.Data.DataTable dataTable = dataTableInput;
            Cursor.Current = Cursors.WaitCursor;
            double[,] dData = new double[dataTable.Rows.Count, dataTable.Columns.Count];
            for (int row = 0; row < dataTable.Rows.Count; row++)
            {
                for (int col = 0; col < dataTable.Columns.Count; col++)
                {
                    dData[row, col] =
```

```
Convert.ToDouble(dataTable.Rows[row].ItemArray[col].ToString());
                    }
                }
                NumericMatrix numericMatrix = rEngine.CreateNumericMatrix(dData);
                rEngine.SetSymbol("df.shp.result", numericMatrix);
                rEngine.Evaluate("df.shp.result <- as.data.frame(df.shp.result)");
                rEngine.Evaluate("names(df.shp.result)[1] <- \"SurvProb\"");
                rEngine.Evaluate("names(df.shp.result)[2] <- \"Edge\"");
                rEngine.Evaluate("names(df.shp.result)[3] <- \"Survival\"");
                rEngine.Evaluate("df.shp.result <- df.shp.result[df.shp.result$Edge == 0, ]");
                rEngine.Evaluate("roc(df.shp.result$Survival, df.shp.result$SurvProb) -> roc.1");
                NumericVector numericVector = rEngine.Evaluate("roc.1$auc").AsNumeric();
                dAUC = Convert.ToDouble(numericVector.First().ToString());
                Cursor.Current = Cursors.Default;
                return dAUC;
            }



            /// <summary>
            /// 分径级显示 AUC;
            /// </summary>
            /// <param name="dataTableInput"></param>
            /// <param name="dgvAUCofDBH"></param>
            /// <returns></returns>
            public bool DBHClassesAUC(System.Data.DataTable dataTableInput, DataGridView dgvAUCofDBH)
            {
                Dictionary<string, double> dictionaryAUCofDBH = new Dictionary<string, double>();
                double dAUC = 0;
                System.Data.DataTable dataTable = dataTableInput;
                Cursor.Current = Cursors.WaitCursor;
                double[,] dData = new double[dataTable.Rows.Count,
```

```csharp
                    dataTable.Columns.Count];
                    for (int row = 0; row < dataTable.Rows.Count; row++)
                    {
                        for (int col = 0; col < dataTable.Columns.Count; col++)
                        {
                            dData[row, col] =
Convert.ToDouble(dataTable.Rows[row].ItemArray[col].ToString());
                        }
                    }
                    NumericMatrix numericMatrix = rEngine.CreateNumericMatrix(dData);
                    rEngine.SetSymbol("df.shp.result", numericMatrix);
                    rEngine.Evaluate("df.shp.result <- as.data.frame(df.shp.result)");
                    rEngine.Evaluate("names(df.shp.result)[1] <- \"SurvProb\"");
                    rEngine.Evaluate("names(df.shp.result)[2] <- \"Edge\"");
                    rEngine.Evaluate("names(df.shp.result)[3] <- \"Survival\"");
                    rEngine.Evaluate("names(df.shp.result)[4] <- \"DBH\"");
                    rEngine.Evaluate("df.shp.result <- df.shp.result[df.shp.result$Edge == 0,]");
                    rEngine.Evaluate("roc(df.shp.result$Survival, df.shp.result$SurvProb) -> roc.1");
                    NumericVector numericVector = rEngine.Evaluate("roc.1$auc").AsNumeric();
                    dAUC = Convert.ToDouble(numericVector.First().ToString());
                    rEngine.Evaluate("roc.data.10 <- df.shp.result[df.shp.result$DBH < 10,]");
                    rEngine.Evaluate("roc.data.20 <- df.shp.result[df.shp.result$DBH >= 10 & df.shp.result$DBH < 20,]");
                    rEngine.Evaluate("roc.data.30 <- df.shp.result[df.shp.result$DBH >= 20 & df.shp.result$DBH < 30,]");
                    rEngine.Evaluate("roc.data.40 <- df.shp.result[df.shp.result$DBH >= 30 & df.shp.result$DBH < 40,]");
                    rEngine.Evaluate("roc.data.50 <- df.shp.result[df.shp.result$DBH >= 40 & df.shp.result$DBH < 50,]");
                    rEngine.Evaluate("roc.data.60 <- df.shp.result[df.shp.result$DBH >= 50 & df.shp.result$DBH < 60,]");
                    rEngine.Evaluate("roc.data.70 <- df.shp.result[df.shp.result$DBH >= 60 & df.shp.result$DBH < 70,]");
```

rEngine.Evaluate ("roc.data.80 <- df.shp.result[df.shp.result$DBH >= 70,]") ;
// AUC;
rEngine.Evaluate ("dbh.10.roc <- roc (roc.data.10$Survival, roc.data.10$SurvProb) ") ;
rEngine.Evaluate ("dbh.20.roc <- roc (roc.data.20$Survival, roc.data.20$SurvProb) ") ;
rEngine.Evaluate ("dbh.30.roc <- roc (roc.data.30$Survival, roc.data.30$SurvProb) ") ;
rEngine.Evaluate ("dbh.40.roc <- roc (roc.data.40$Survival, roc.data.40$SurvProb) ") ;
rEngine.Evaluate ("dbh.50.roc <- roc (roc.data.50$Survival, roc.data.50$SurvProb) ") ;
rEngine.Evaluate ("dbh.60.roc <- roc (roc.data.60$Survival, roc.data.60$SurvProb) ") ;
rEngine.Evaluate ("dbh.70.roc <- roc (roc.data.70$Survival, roc.data.70$SurvProb) ") ;
rEngine.Evaluate ("dbh.80.roc <- roc (roc.data.80$Survival, roc.data.80$SurvProb) ") ;
NumericVector numericVectorDBH10 = rEngine.Evaluate ("dbh.10.roc$auc") .AsNumeric () ;
double dAUC10 = Convert.ToDouble (numericVectorDBH10.First () .ToString ()) ;
dictionaryAUCofDBH.Add ("DBH: < 10 cm", dAUC10) ;
NumericVector numericVectorDBH20 = rEngine.Evaluate ("dbh.20.roc$auc") .AsNumeric () ;
double dAUC20 = Convert.ToDouble (numericVectorDBH20.First () .ToString ()) ;
dictionaryAUCofDBH.Add ("DBH: 10 ~ 20 cm", dAUC20) ;
NumericVector numericVectorDBH30 = rEngine.Evaluate ("dbh.30.roc$auc") .AsNumeric () ;
double dAUC30 = Convert.ToDouble (numericVectorDBH30.First () .ToString ()) ;
dictionaryAUCofDBH.Add ("DBH: 20 ~ 30 cm", dAUC30) ;
NumericVector numericVectorDBH40 = rEngine.Evaluate ("dbh.40.roc$auc") .AsNumeric () ;

```csharp
            double dAUC40 =
Convert.ToDouble(numericVectorDBH40.First().ToString());
            dictionaryAUCofDBH.Add("DBH: 30 ~ 40 cm", dAUC40);
            NumericVector numericVectorDBH50 =
rEngine.Evaluate("dbh.50.roc$auc").AsNumeric();
            double dAUC50 =
Convert.ToDouble(numericVectorDBH50.First().ToString());
            dictionaryAUCofDBH.Add("DBH: 40 ~ 50 cm", dAUC50);
            NumericVector numericVectorDBH60 =
rEngine.Evaluate("dbh.60.roc$auc").AsNumeric();
            double dAUC60 =
Convert.ToDouble(numericVectorDBH60.First().ToString());
            dictionaryAUCofDBH.Add("DBH: 50 ~ 60 cm", dAUC60);
            NumericVector numericVectorDBH70 =
rEngine.Evaluate("dbh.70.roc$auc").AsNumeric();
            double dAUC70 =
Convert.ToDouble(numericVectorDBH70.First().ToString());
            dictionaryAUCofDBH.Add("DBH: 60 ~ 70 cm", dAUC70);
            NumericVector numericVectorDBH80 =
rEngine.Evaluate("dbh.80.roc$auc").AsNumeric();
            double dAUC80 =
Convert.ToDouble(numericVectorDBH80.First().ToString());
            dictionaryAUCofDBH.Add("DBH: > 70 cm", dAUC80);
            DataGridViewTextBoxColumn acDBH =
new DataGridViewTextBoxColumn();
            acDBH.Name = "DBH";
            acDBH.DataPropertyName = "DBH";
            acDBH.HeaderText = "DBH";
            dgvAUCofDBH.Columns.Add(acDBH);
            DataGridViewTextBoxColumn acAUC =
new DataGridViewTextBoxColumn();
            acAUC.Name = "AUC";
            acAUC.DataPropertyName = "AUC";
            acAUC.HeaderText = "AUC";
            dgvAUCofDBH.Columns.Add(acAUC);
```

```csharp
foreach (KeyValuePair<string, double> kvp in dictionaryAUCofDBH)
{
    int index = dgvAUCofDBH.Rows.Add();
    dgvAUCofDBH.Rows[index].Cells[0].Value = kvp.Key.ToString();
    dgvAUCofDBH.Rows[index].Cells[1].Value = kvp.Value;
}
dgvAUCofDBH.Refresh();
Cursor.Current = Cursors.Default;
return true;
}


/// <summary>
/// 空间分布图;
/// </summary>
/// <param name="dataTable"></param>
public void SpatialDistributionMap(System.Data.DataTable dataTable, PictureBox pictureBox)
{
    Cursor.Current = Cursors.WaitCursor;
    double[,] dData = new double[dataTable.Rows.Count, dataTable.Columns.Count];
    for (int row = 0; row < dataTable.Rows.Count; row++)
    {
        for (int col = 0; col < dataTable.Columns.Count; col++)
        {
            dData[row, col] = Convert.ToDouble(dataTable.Rows[row].ItemArray[col].ToString());
        }
    }
    NumericMatrix numericMatrix = rEngine.CreateNumericMatrix(dData);
    rEngine.SetSymbol("df.shp.result", numericMatrix);
    rEngine.Evaluate("setwd(\"D:/\")");
    rEngine.Evaluate("df.shp.result <- as.data.frame(df.shp.result)");
    rEngine.Evaluate("names(df.shp.result)[1] <- \"SurvProb\"");
```

```csharp
rEngine.Evaluate ("names (df.shp.result) [2] <- \"gx\"");
rEngine.Evaluate ("names (df.shp.result) [3] <- \"gy\"");
rEngine.Evaluate ("names (df.shp.result) [4] <- \"Edge\"");
rEngine.Evaluate ("df.shp.result <- df.shp.result[df.shp.result$Edge == 0, ]");
rEngine.Evaluate ("names (df.shp.result) [2] <- \"x\"");
rEngine.Evaluate ("names (df.shp.result) [3] <- \"y\"");
rEngine.Evaluate ("cuts <- seq (0, 1, length.out = 8) ");
rEngine.Evaluate ("cuts <- round (cuts, digits = 5) ");
rEngine.Evaluate ("col.regions <- brewer.pal (length (cuts) + 3 - 1, \"RdYlGn\") ");
rEngine.Evaluate ("ggplot (data = df.shp.result, aes (x = x, y = y)) + geom_point (aes (colour = SurvProb), size = 1) + scale_colour_gradientn (colours = col.regions) + coord_cartesian (xlim = c (0, 300)) + guides (fill = FALSE) + coord_cartesian (ylim = c (0, 300)) + xlab (\"West-East\") + ylab (\"South-North\") + theme (panel.grid.major = element_blank ()) + theme (panel.grid.minor = element_blank ()) + coord_equal () ");
rEngine.Evaluate ("ggsave (file=\"spatial.plot.jpg\") ");
string strProgramPath = System.Windows.Forms.Application.StartupPath + "\\Image\\spatial.plot.jpg";
string strSpatialDistributionMap = "D:\\spatial.plot.jpg";
if (File.Exists (strProgramPath))
{
    File.SetAttributes (strProgramPath, FileAttributes.Normal);
}
File.Copy (strSpatialDistributionMap, strProgramPath, true);
pictureBox.SizeMode = PictureBoxSizeMode.Zoom;
pictureBox.Image = Image.FromFile (strProgramPath);
System.IO.File.Delete (strSpatialDistributionMap);
}

/// <summary>
/// 光滑密度分布图；
/// </summary>
/// <param name="dataTable"></param>
```

```csharp
public void SmoothDensityPlot (System.Data.DataTable dataTable, PictureBox pictureBox)
{
    Cursor.Current = Cursors.WaitCursor;
    double[,] dData = new double[dataTable.Rows.Count, dataTable.Columns.Count];
    for (int row = 0; row < dataTable.Rows.Count; row++)
    {
        for (int col = 0; col < dataTable.Columns.Count; col++)
        {
            dData[row, col] = Convert.ToDouble (dataTable.Rows[row].ItemArray[col].ToString ());
        }
    }
    NumericMatrix numericMatrix = rEngine.CreateNumericMatrix (dData);
    rEngine.SetSymbol ("df.shp.result", numericMatrix);
    rEngine.Evaluate ("setwd (\"D:/\")");
    rEngine.Evaluate ("df.shp.result <- as.data.frame (df.shp.result)");
    rEngine.Evaluate ("names (df.shp.result) [1] <- \"SurvProb\"");
    rEngine.Evaluate ("names (df.shp.result) [2] <- \"Edge\"");
    rEngine.Evaluate ("names (df.shp.result) [3] <- \"Survival\"");
    rEngine.Evaluate ("df.shp.result <- df.shp.result[df.shp.result$Edge == 0, ]");
    rEngine.Evaluate ("ggplot (data = df.shp.result, aes (SurvProb, fill = \"blue\")) + geom_density (alpha = I(0.6)) + coord_cartesian (xlim = c (0, 1)) + guides (fill = FALSE) + xlab (\"Predicted Tree Survival Probability\") + ylab (\"Density\") + theme (panel.grid.major = element_blank ()) + theme (panel.grid.minor = element_blank ())");
    rEngine.Evaluate ("ggsave (file=\"density.plot.jpg\")");
    string strProgramPath = System.Windows.Forms.Application.StartupPath + "\\Image\\density.plot.jpg";
    string strSmoothDensityPlot = "D:\\density.plot.jpg";
    System.IO.File.Copy (strSmoothDensityPlot, strProgramPath, true);
    pictureBox.SizeMode = PictureBoxSizeMode.Zoom;
    pictureBox.Image = Image.FromFile (strProgramPath);
    System.IO.File.Delete (strSmoothDensityPlot);
```

        }

    }

    /// <summary>
    /// 非线性回归接口;
    /// </summary>
    public interface INonLinearRegression
    {
        void PowerFunction(ref double dA, ref double dB, ref double dIntercept, DataTable dataTable);
        void SingleMolecularEquation(ref double dHmax, ref double dS, DataTable dataTable);
    }

    /// <summary>
    /// 线性回归接口;
    /// </summary>
    public interface ILinearRegression
    {
        void UnivariateLinearModel(out double dIntercept, out double dSlope, DataTable dataTable);
    }

    /// <summary>
    /// 树种参数估计类;
    /// </summary>
    public class SpeciesParameterEstimation : INonLinearRegression, ILinearRegression
    {
        REngine rEngine = null;
        public SpeciesParameterEstimation()
        {
            REngine.SetEnvironmentVariables();
            rEngine = REngine.GetInstance();

```csharp
/// <summary>
/// 幂函数(树高曲线；叶面积；树干生物量等);
/// </summary>
/// <param name="dA"></param>
/// <param name="dB"></param>
/// <param name="dIntercept"></param>
/// <param name="dataTable"></param>
public void PowerFunction(ref double dA, ref double dB, ref double dIntercept, DataTable dataTable)
{
    DataFrame dataFrame = CreateDataFrame2Column(dataTable);
    rEngine.SetSymbol("df.data", dataFrame);
    rEngine.Evaluate("names(df.data)[1] <- \"Y\"");
    rEngine.Evaluate("lm.fit <- lm(log(Y) ~ log(X), data = df.data)");
    if (dIntercept == 0)
    {
        rEngine.Evaluate("st <- list(a = exp(coef(lm.fit)[1]), b = coef(lm.fit)[2])");
        rEngine.Evaluate("fit <- nls(Y ~ a * (X ^ b), start = st, data = df.data)");
        dIntercept = 0;
    }
    else
    {
        rEngine.Evaluate("st <- list(a = exp(coef(lm.fit)[1]), b = coef(lm.fit)[2], c = 1.3)");
        rEngine.Evaluate("fit <- nls(Y ~ c + a * (X ^ b), start = st, data = df.data)");
    }
    rEngine.Evaluate("coef(fit)[1] -> a");
    rEngine.Evaluate("coef(fit)[2] -> b");
    NumericVector numericVectorA = rEngine.Evaluate("a").AsNumeric();
    dA = Convert.ToDouble(numericVectorA.First().ToString());
    NumericVector numericVectorB = rEngine.Evaluate("b").AsNumeric();
```

```csharp
            dB = Convert.ToDouble(numericVectorB.First().ToString());
        }


        //
        /// <summary>
        /// 单分子模型(树高曲线);
        /// </summary>
        /// <param name="dDBH0"></param>
        /// <param name="dHmax"></param>
        /// <param name="dS"></param>
        /// <param name="dataTable"></param>
        public void SingleMolecularEquation(ref double dHmax, ref double dS, DataTable dataTable)
        {
            DataFrame dataFrame = CreateDataFrame2Column(dataTable);
            rEngine.SetSymbol("df.data", dataFrame);
            rEngine.Evaluate("names(df.data)[1] <- \"Y\"");
            rEngine.Evaluate("names(df.data)[2] <- \"X\"");
            NumericVector numericVectorS = rEngine.CreateNumeric(dS);
            NumericVector numericVectorHmax = rEngine.CreateNumeric(dHmax);
            rEngine.SetSymbol("s.initial", numericVectorS);
            rEngine.SetSymbol("Hmax.initial", numericVectorHmax);
            rEngine.Evaluate("st <- list(s = s.initial, Hmax = Hmax.initial)");
            rEngine.Evaluate("fit <- nls(Y ~ 1.3 + (Hmax-1.3) * (1 - exp(-s * X / (Hmax - 1.3))), start = st, data = df.data)");
            rEngine.Evaluate("coef(fit)[1] -> a");
            rEngine.Evaluate("coef(fit)[2] -> b");
            NumericVector numericVectorA = rEngine.Evaluate("a").AsNumeric();
            dS = Convert.ToDouble(numericVectorA.First().ToString());
            NumericVector numericVectorB = rEngine.Evaluate("b").AsNumeric();
            dHmax = Convert.ToDouble(numericVectorB.First().ToString());

        }
```

```csharp
//
/// <summary>
/// 一元线性回归;
/// </summary>
/// <param name="dIntercept"></param>
/// <param name="dSlope"></param>
/// <param name="dataTable"></param>
public void UnivariateLinearModel(out double dIntercept, out double dSlope, DataTable dataTable)
{
    DataFrame dataFrame = CreateDataFrame2Column(dataTable);
    rEngine.SetSymbol("df.data", dataFrame);
    rEngine.Evaluate("names(df.data)[1] <- \"Y\"");
    rEngine.Evaluate("names(df.data)[2] <- \"X\"");
    rEngine.Evaluate("lm.fit <- lm(Y ~ X, data = df.data)");
    rEngine.Evaluate("coef(lm.fit)[1] -> a");
    rEngine.Evaluate("coef(lm.fit)[2] -> b");
    NumericVector numericVectorA = rEngine.Evaluate("a").AsNumeric();
    dIntercept = Convert.ToDouble(numericVectorA.First().ToString());
    NumericVector numericVectorB = rEngine.Evaluate("b").AsNumeric();
    dSlope = Convert.ToDouble(numericVectorB.First().ToString());
}


/// <summary>
/// 创建 DataFrame 两列数据;
/// </summary>
/// <param name="dataTable"></param>
/// <returns></returns>
private DataFrame CreateDataFrame2Column(DataTable dataTable)
{
    IEnumerable[] columns = new IEnumerable[2];
    double[,] dData = new double[dataTable.Rows.Count, dataTable.Columns.Count];
    for (int col = 0; col < dataTable.Columns.Count; col++)
```

```csharp
            {
                columns[col] = new double[] { };
                double[] dValues = new double[dataTable.Rows.Count];
                for (int row = 0; row < dataTable.Rows.Count; row++)
                {
                    double dValue = Convert.ToDouble(dataTable.Rows[row].ItemArray[col].ToString());
                    dValues[row] = dValue;
                }
                columns[col] = dValues;
            }
            string[] columnNames = new[] { "Y", "X" };
            DataFrame dataFrame = rEngine.CreateDataFrame(columns, columnNames);
            return dataFrame;
        }

    }

}
```